T0200256

HANDBOOKS
STARS & PLANETS

DK HANDBOOKS

STARS & PLANETS

IAN RIDPATH

Star charts by
Royal Greenwich Observatory

Important Notice
Observing the Sun through any kind of optical instrument can cause blindness.
The author and publishers cannot accept any responsibility for readers
who ignore this advice.

NEW EDITION

DK LONDON

Managing Editor Angeles Gavira Guerrero
Managing Art Editor Michael Duffy
Production Editor Gillian Reid
Senior Production Controller Meskerem Berhane
Jacket Design Development Manager Sophia MTT
Associate Publishing Director Liz Wheeler
Art Director Karen Self
Publishing Director Jonathan Metcalf

DK DELHI

Desk Editor Saumya Agarwal
Project Art Editor Rupanki Arora Kaushik
Managing Editor Saloni Singh
Managing Art Editor Sudakshina Basu
Jacket Designer Juhi Sheth
Senior DTP Designers Pushpak Tyagi
DTP Designers Anurag Trivedi, Rakesh Kumar
Pre-production Manager Sunil Sharma
Editorial Head Glenda Fernandes
Design Head Malavika Talukder

Author Ian Ridpath
Consultants Giles Sparrow, Robin Scagell

FIRST EDITION

Project Editor Peter Frances
DTP Designer Rob Campbell
Project Designer Helen Taylor
Picture Research Angela Anderson, Robin Scagell
Production Controllers Michelle Thomas, Alison Jones

This edition published in 2022
First published in Great Britain in 1998 by
Dorling Kindersley Limited
DK, One Embassy Gardens, 8 Viaduct Gardens,
London, SW11 7BW

The authorised representative in the EEA is
Dorling Kindersley Verlag GmbH. Arnulfstr. 124,
80636 Munich, Germany

A CIP catalogue record for this book
is available from the British Library.
ISBN: 978-0-2415-5856-0

Printed and bound in China

ww.dk.com

This book was made with Forest
Stewardship Council™ certified
paper – one small step in DK's
commitment to a sustainable future.
Learn more at
www.dk.com/uk/information/sustainability

Contents

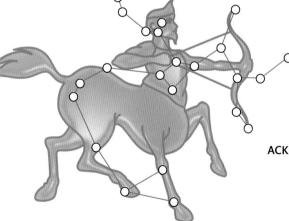

AUTHOR'S INTRODUCTION

AS THE SUN SETS and the sky darkens, the marvels of the night sky come into view: planets in our Solar System; stars, star clusters, and nebulae in our Galaxy; and other galaxies, unimaginably distant. Examples of all these objects can be seen with the naked eye, while binoculars or a small telescope reveal many more. This book shows you where and how to find them.

Astronomy is usually regarded as the oldest of the sciences – yet, with the continuing development of space probes to the planets and orbiting telescopes, it is also among the most modern.

ANCIENT ASTRONOMY

Serious study of the sky began thousands of years ago in the Middle East and, at least in the ancient world, reached a pinnacle with the Greeks 2,000 years ago. At that time, the stars and planets were still mysterious lights in the sky, and the Earth was thought to be at the centre of the Universe. This view was not seriously challenged until the 16th century, when Nicolaus Copernicus, a Polish astronomer, proposed that the Earth is simply a planet and that all planets orbit the Sun. His revolutionary view was confirmed in the following century by the Italian Galileo Galilei, using the newly invented telescope. Johannes Kepler, a German mathematician, worked out that planets orbit on elliptical paths, and the Englishman Isaac Newton explained the gravitational forces governing all orbital motions.

MODERN ASTRONOMY

From Newton's time in the 17th century onwards, it was recognized that the stars are other suns, but only in the 20th century, with the work of the American astronomer Edwin Hubble, did it become clear that our Galaxy of stars is only one among countless others, and that the Universe as a whole is expanding, as though from a massive explosion billions of years ago. And only with the development of nuclear physics has it become clear how the stars generate the energy that makes them shine.

SKYWATCHING

For all its modern advances, astronomy remains a branch of science where amateurs can make a real contribution, often with modest equipment – for example, monitoring changes in the brightness of variable stars, logging meteor showers, following storms in the atmospheres of Mars, Jupiter, and Saturn – and, if you're lucky, discovering an exploding star (a nova) or a comet. Perhaps one of the readers of this book will go on to make such a discovery.

The Farnese Atlas
The celestial globe held here by Atlas shows the constellations known to the ancient Greeks.

HOW THIS BOOK WORKS

THIS BOOK is divided into four parts: a general introduction to astronomy; a detailed guide to the Solar System; an alphabetical catalogue of the constellations; and a month-by-month companion to the night sky. Sample pages from the three main sections (excluding the Introduction) are shown below. Further explanation appears at the start of each section.

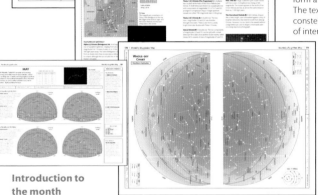

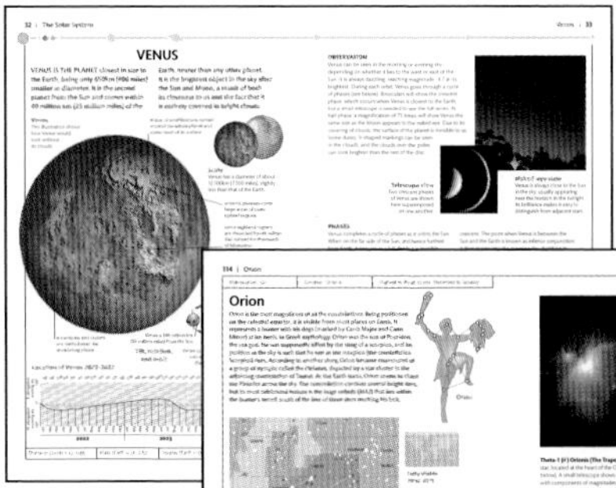

The Solar System, pp.25–62
This section contains descriptions of the eight planets, the Sun and Moon, and the more significant smaller bodies. It also includes advice on how and where to find them.

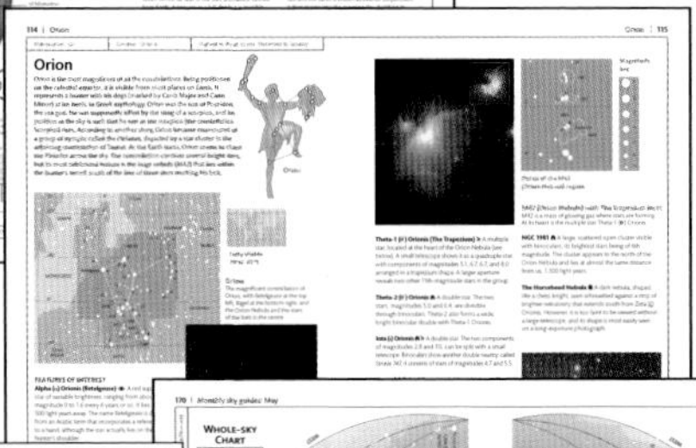

The constellations, pp.63–141
Entries on the 88 constellations are arranged alphabetically by name. Collectively, the charts in this section form a complete atlas of the sky. The text describes the origins of the constellations and selected features of interest.

Monthly sky guide, pp.142–215
An introduction to each month, with easy-to-use charts and descriptions of what to look out for, is followed by more detailed charts of the night sky as seen from northern and southern latitudes.

Introduction to the month

KEY TO SYMBOLS

Greek alphabet
Greek letters form part of the names of certain stars and appear on some of the charts in this book.

α	alpha	ι	iota	ρ	rho
β	beta	κ	kappa	σ	sigma
γ	gamma	λ	lambda	τ	tau
δ	delta	μ	mu	υ	upsilon
ε	epsilon	ν	nu	φ	phi
ζ	zeta	ξ	xi	χ	chi
η	eta	ο	omicron	ψ	psi
θ	theta	π	pi	ω	omega

Deep-sky objects
These symbols appear in red on the constellation and whole-sky charts.

- ◯ Galaxy
- ⊛ Globular cluster
- ⁘ Open cluster
- ☐ Diffuse nebula
- (◯) Planetary nebula
- △ X-ray and radio sources

Observation symbols
These symbols indicate the type of instrument needed to observe features described in the constellations section.

- ◉ Naked eye
- � Binoculars
- ⚞ Telescope
- ♟ Not observable with amateur equipment

THE UNIVERSE

THE UNIVERSE is everything that exists – all matter, space, and time. It extends as far as the largest telescopes can see, at least 10,000 million light years in all directions around us. The Universe is thought to have formed in a gigantic explosion, the Big Bang, some 13.8 billion years ago and is still expanding. Within the Universe, matter is held together by gravity in structures of varying size.

OUR GALAXY

Most of the visible matter in the Universe is grouped in huge aggregations of stars, gas, and dust called galaxies. Our nearest star, the Sun, is an insignificant star in a galaxy of at least 100,000 million others (usually known as "the Galaxy"). Although it is difficult to tell from our position inside it, the Galaxy is thought to be spiral in shape. There is some evidence indicating that there is a bar of material across the centre. The Galaxy is about 100,000 light years in diameter, and we are located about two-thirds of the way from the centre to the rim, in one of its spiral arms. The stars that we see in the sky are all part of the Galaxy. Those nearest to us appear in all directions in the sky, and form the constellations (see p.18). Since the Galaxy is flattened in shape, more distant stars mass into a hazy band known as the Milky Way.

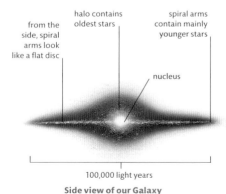

The Milky Way

The Milky Way, a pale band that can be seen arching across the sky on dark nights, consists of innumerable distant stars within our Galaxy. In this view of the Milky Way, looking towards the centre of the Galaxy, apparent gaps are caused by clouds of dark dust that lie in the spiral arms, obscuring the light from stars behind.

stars orbit the centre of the Galaxy

outer stars take longer to orbit the centre than inner stars

Perseus Arm

bright nucleus

outer arm

Solar System is located in Orion (Local) Arm

Sagittarius Arm

Overhead view of our Galaxy

from the side, spiral arms look like a flat disc

halo contains oldest stars

spiral arms contain mainly younger stars

nucleus

100,000 light years

Side view of our Galaxy

OUR GALAXY

The Universe contains countless galaxies, some of which we can see from the Earth. Galaxies are divided into spirals, barred spirals, ellipticals, and irregulars. Spirals have arms which consist of relatively young stars, gas clouds, and dust, and which curve out from a central bulge of old stars. In barred spirals, the arms extend from the ends of a central bar. Elliptical galaxies, consisting of old stars, have no arms and very little gas or dust. Irregular galaxies have no regular shape or structure. The largest galaxies are ellipticals, with more than ten times the mass of our Galaxy; some giant ellipticals may have formed by the merger of smaller galaxies. Far off in the Universe are objects called quasars that emit as much energy as a galaxy from an area not much larger than the Solar System. Quasars, and their less luminous relatives, known as Seyfert galaxies and BL Lacertae objects, are thought to be galaxies with massive black holes at their centres.

Spiral galaxy
This galaxy, M83, appears "face-on" to us, so that we see its nucleus and curving arms.

Barred spiral galaxy
In a barred spiral, such as NGC 1365, a bar of stars and gas lies across the nucleus.

Elliptical galaxy
Elliptical galaxies range from dwarfs to giants. M87 is an example of a giant elliptical.

GROUPS OF GALAXIES

Galaxies usually occur in groups known as clusters. Our Galaxy is the second largest in a relatively small family of over 50 galaxies called the Local Group, about 3 million light years across. The Group's largest member is the Andromeda Galaxy, a spiral galaxy about 2.5 million light years away and the most distant object visible to the naked eye. Some clusters have thousands of members, as does the Virgo Cluster, the nearest large cluster to us, about 55 million light years away. Its brightest members can be seen with the size of telescope used by amateur astronomers.

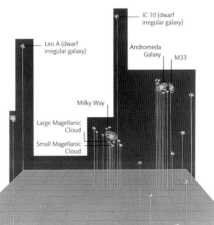

IC 10 (dwarf irregular galaxy)

Leo A (dwarf irregular galaxy)

Andromeda Galaxy

M33

Milky Way

Large Magellanic Cloud

Small Magellanic Cloud

The Local Group

WHAT IS A STAR?

STARS ARE BALLS of gas that release energy produced by nuclear reactions within their cores. Most stars are similar to the Sun, our nearest star, but because they are so far away they appear to us as mere points of light. Throughout the Galaxy, stars are forming, evolving, and being destroyed. By studying a range of stars, astronomers have been able to build a picture of how stars change over time, and hence understand more about the past and probable future of the Sun.

STAR FORMATION

Stars form in huge clouds of gas and dust in space called nebulae, in a process that is continuing today. The nebula shrinks under the inward pull of its own gravity, forming an embryonic star known as a protostar. Eventually, the density and temperature of the gas at the protostar's centre become high enough for nuclear reactions to begin. The object "switches on" to become a true star, generating its own heat and light. The star is then said to be on the main sequence. How long it stays like this and what happens next depend on the star's mass.

Star formation

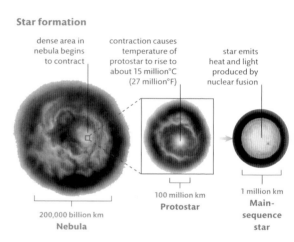

dense area in nebula begins to contract

contraction causes temperature of protostar to rise to about 15 million°C (27 million°F)

star emits heat and light produced by nuclear fusion

200,000 billion km
Nebula

100 million km
Protostar

1 million km
Main-sequence star

NEBULAE

Nebulae are clouds of gas and dust within galaxies. Bright diffuse nebulae are areas of hydrogen gas where new stars are being formed. Two types of nebula are linked with the late stages of star development: planetary nebulae are shells of gas thrown off by red giant stars, while supernova remnants are the twisted wreckage of exploded massive stars. Some nebulae are dark because they contain no stars to illuminate them, and are visible only when silhouetted against a brighter background.

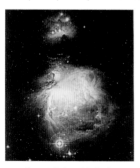

**Great Orion Nebula
(Bright diffuse nebula)**

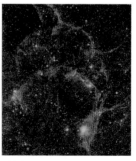

**Vela supernova remnant
(Supernova remnant)**

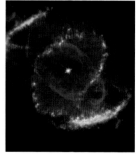

**Cat's Eye Nebula
(Planetary nebula)**

MASSIVE STARS

A main-sequence star with a mass of more than about 10 Suns experiences a spectacular end. It swells into a red supergiant with cooling, expanding outer layers. Eventually its core collapses, causing a huge explosion known as a supernova. For a few weeks, the supernova shines as brightly as an entire galaxy. While the outer layers of the star are scattered in space, the fate of the core again depends on its mass. A core of relatively low mass will be crushed into a tiny, superdense neutron star. If the core has a mass of more than about two Suns, its own gravity will squash it further, into a black hole.

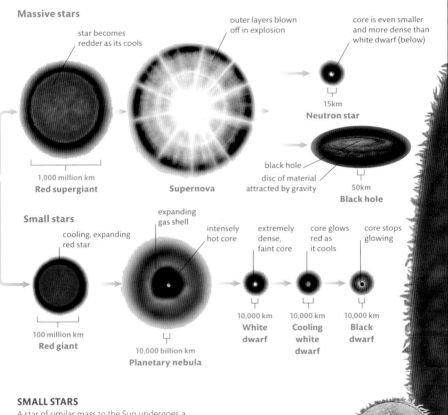

Massive stars

star becomes redder as its cools

outer layers blown off in explosion

core is even smaller and more dense than white dwarf (below)

1,000 million km
Red supergiant

Supernova

15km
Neutron star

black hole
disc of material attracted by gravity

50km
Black hole

Small stars

cooling, expanding red star

expanding gas shell

intensely hot core

extremely dense, faint core

core glows red as it cools

core stops glowing

100 million km
Red giant

10,000 billion km
Planetary nebula

10,000 km
White dwarf

10,000 km
Cooling white dwarf

10,000 km
Black dwarf

SMALL STARS

A star of similar mass to the Sun undergoes a quieter end than a massive star. It swells into a red giant, eventually losing its outer layers, which form a gas shell known as a planetary nebula (because of its resemblance to the disc of a distant planet). The core is exposed as a hot white dwarf that cools and fades over billions of years. On ceasing to emit light, it will become a black dwarf. The least massive stars, red dwarfs with only about one-tenth the Sun's mass, can last for 100 billion years or more, whereas the most massive stars burn out after only about a million years. The Sun, which formed about 4,600 million years ago, is about halfway through its life-cycle.

Star sizes
Stars vary greatly in size. A red giant may have a diameter of up to 150 million km (100 million miles). The Sun (a main-sequence star) is 1.39 million km (865,000 miles) in diameter, while a white dwarf, the size of which is exaggerated on this diagram, is typically only 10,000km (6,000 miles) across.

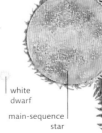

white dwarf

main-sequence star

red giant

STAR FAMILIES AND VARIABLE STARS

STARS ARE not always the simple, unchanging points of light that they may at first appear to be. Many of them occur in families of two or three, or sometimes in much larger groupings. Others vary in brightness in cycles that last days, months, or years. Examples of double and multiple stars, star clusters, and variable stars can be seen with small instruments or even the naked eye.

DOUBLE AND MULTIPLE STARS

When seen with binoculars or a telescope, many stars turn out to have one or more companions. Sometimes the companion stars are unrelated objects in the foreground or background, in which case the pair form what is called an optical double. But in most cases, the stars lie close together in space, forming a pairing known as a binary. Linked by gravity, the members of a binary orbit around each other, although it usually takes many years to detect any motion, and in the case of multiple stars the orbital motions can be very complex. Separating the components of double stars is a popular activity among amateur astronomers. The closer together the stars are as seen from Earth, the larger the aperture of telescope needed to divide (or "split") them.

Albireo
A colour contrast of orange and blue makes this a particularly attractive double star for users of small telescopes.

When the stars are very unequal in brightness, the fainter star (the secondary) can be difficult to see because of the glare from the brighter one (the primary). Some doubles, called spectroscopic binaries, have components that are too close together to be separated through an optical telescope. Only by studying the spectra of light from these stars are professional astronomers able to establish that they are double.

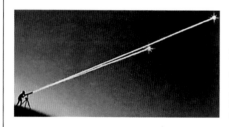

Optical double stars
Two stars that lie in the same line of sight but are at different distances from us are known as an optical double. Double stars of this type are less common than genuine binaries. In fact, more than three-quarters of all stars are linked to one or more companions by gravity.

STARS LINKED BY GRAVITY

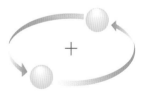

Equal mass
In a binary star with members of equal mass, the stars orbit around a centre of mass midway between them.

Unequal mass
When one member of a binary is more massive than the other, the centre of mass lies closer to the heavier star.

A multiple star
Here, four stars of equal mass form two pairs orbiting a centre of mass. Multiple star orbits are usually highly elliptical.

STAR CLUSTERS

Rather than originating singly, most stars form as part of a group or cluster, of which there are two distinct types. Open clusters, the more common type, have no specific shape, being scattered collections of dozens to hundreds of relatively young stars lying in the spiral arms of our Galaxy. Globular clusters, which lie in a halo around our Galaxy, are spherical or slightly elliptical in shape and contain large numbers of very old stars.

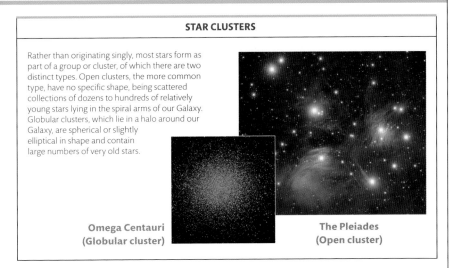

**Omega Centauri
(Globular cluster)**

**The Pleiades
(Open cluster)**

VARIABLE STARS

A variable star is one that varies in brightness over time. Nearly 20 per cent of such stars are eclipsing binaries – very close pairs in which one star periodically passes in front of the other, causing a drop in the total amount of light reaching the Earth. The most famous eclipsing binary is a star called Algol (in the constellation Perseus). But most stars that vary in brightness do so because they pulsate in size. The most common of these are red giants and supergiants called Mira stars, after their prototype, Mira (in Cetus). Such stars vary by up to 25,000 times in brightness over three months to three years. Many other red giants and supergiants also pulsate, but with less regularity and smaller changes in output than Mira stars. Relatively rare, but important, are Cepheid variables, named after Delta Cephei (in Cepheus). The pulsation period of these yellow supergiant stars is directly related to their luminosity, so by timing a star's variations astronomers can work out its actual brightness. Since the brightness of a star in our sky depends on its distance from us, such stars act as valuable "standard candles" for measuring distances in space.

Brightness of Algol and Delta Cephei

These graphs show how the apparent brightness (or magnitude) of these two stars varies over time. While the curve for Delta Cephei is relatively smooth, the brightness of Algol falls dramatically when the primary is eclipsed by its fainter partner.

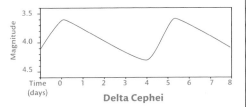

Delta Cephei

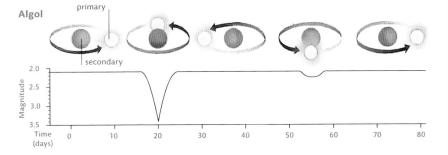

Algol

THE SOLAR SYSTEM

THE HUB of the Solar System is the Sun. Around it orbit eight major planets and their moons, along with a swarm of asteroids (also known as minor planets) and several dwarf planets. The Sun contains about 99.9 per cent of the Solar System's total mass, and all the objects in the system are held in place by its gravity. At the edge of the Solar System is a cloud of comets that extends roughly halfway to the nearest star, a distance of about two light years.

THE SUN AND PLANETS

The Solar System is orbiting the centre of the Galaxy. At the same time, the planets are in orbit around the Sun, all moving anti-clockwise as viewed from over the Sun's north pole. Each planet also spins on its own axis of rotation. The four innermost planets – Mercury, Venus, Earth, and Mars – are small, rocky bodies, sometimes known as the terrestrial planets. Beyond Mars is the asteroid belt and then the gas giants: Jupiter, Saturn, Uranus, and Neptune, so-called because they are much larger than the Earth and consist mostly of hydrogen and helium. All four have ring systems and numerous moons. Beyond Neptune lies a belt of small, icy objects that includes the dwarf planets Pluto and Eris. The planets are thought to have formed from a disc of gas and dust around the newly formed Sun about 4.6 billion years ago. Astronomers have found evidence that planets exist around other stars, too.

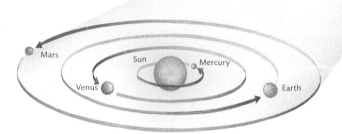

Uranus

Orbits of the inner planets

The Earth is the largest of the four rocky inner planets. Mercury and Venus, which orbit between the Sun and Earth, are known as inferior planets.

Mars

Sun

Mercury

Venus

Earth

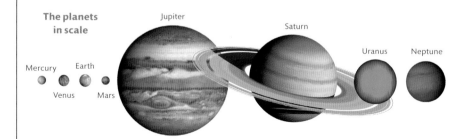

The planets in scale

Jupiter

Saturn

Uranus

Neptune

Mercury

Earth

Venus

Mars

OBSERVING THE PLANETS

The orbits of the Earth and other planets are in much the same plane. As a result, the planets stay close to an imaginary line called the ecliptic (see p.16) that represents the Sun's apparent path across the sky over the course of a year. Inferior planets (Mercury and Venus) are always close to the Sun in the sky, appearing before dawn or after sunset, but never in a truly dark sky. Five planets – Mercury, Venus, Mars, Jupiter, and Saturn – are bright enough to be seen with the naked eye. They can be identified by watching their movement from night to night against the backdrop of seemingly fixed stars, noting how they disturb the shapes of the constellations. The outermost planets are harder to see: Uranus and Neptune can be found with binoculars, using the finder charts on p.55 and p.57.

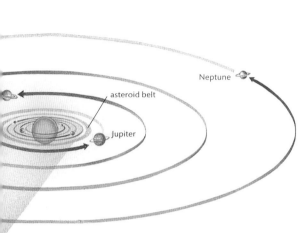

apparent path of Mars

Mars

Neptune

asteroid belt

Jupiter

Earth   Sun

Retrograde Mars

Orbits of the outer planets
The five planets beyond the Earth are referred to as superior planets. Like all planets, their orbits take longer to complete the further they are from the Sun.

Retrograde motion
Relative to the stars, the planets seem to move across the Earth's sky from west to east. But when the Earth, which is travelling more quickly on its inside track around the Sun, overtakes a superior planet, the planet seems to execute a backward loop in the sky. Known as retrograde motion, this effect is most obvious in the motion of Mars.

OPPOSITION AND CONJUNCTION

The ease with which most planets can be seen depends on their position in relation to the Sun and Earth. The angle between the Sun and a planet, as seen from the Earth, is known as elongation. At their greatest eastern and western elongations, inferior planets lie in the evening and morning skies respectively. At conjunction (when elongation is zero), a planet is lost in the Sun's glare. A superior planet can lie opposite the Sun in the sky, a position called opposition; at this time, it looks largest, is visible all night, and at midnight lies due south from northern latitudes and due north from southern latitudes.

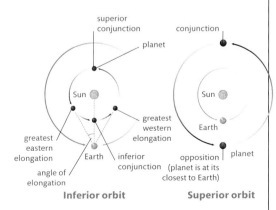

superior conjunction

conjunction

planet

Sun

Sun

Earth

greatest western elongation

greatest eastern elongation

Earth   inferior conjunction

opposition   planet
(planet is at its closest to Earth)

angle of elongation

Inferior orbit

Superior orbit

THE CELESTIAL SPHERE

ALL CELESTIAL OBJECTS appear to be attached to an invisible sphere of infinite size centred on the Earth. Known as the celestial sphere, this enormous imaginary globe seems to spin around the Earth once a day, although it is actually the Earth that is rotating. The portion of the celestial sphere seen from the Earth depends on the observer's latitude, the time of night, and the time of year.

THE CELESTIAL SPHERE

There are various important points and lines on the celestial sphere, similar to those on the Earth's globe. Directly over the Earth's poles lie the celestial poles, around which the sphere appears to turn each day. The celestial equator is a circle on the celestial sphere above the Earth's equator. Another circle, known as the ecliptic, represents the Sun's apparent path around the sphere each year. Of course, the Sun's motion is actually due to the Earth orbiting the Sun; hence the ecliptic is really the plane of the Earth's orbit projected on to the celestial sphere. Because the Earth's axis is tilted at 23.5 degrees, the celestial equator is tilted at the same angle to the ecliptic.

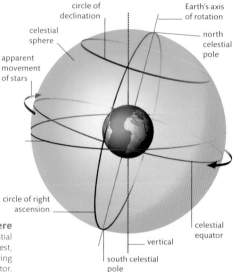

The celestial sphere
Viewed from the Earth, the celestial sphere appears to turn from east to west, with the stars and other objects moving parallel to the celestial equator.

LATITUDE

How much of the celestial sphere we can see depends on our latitude. From either pole, only half the sky can ever be seen; on any night, objects circle the celestial pole but do not rise or set (they are circumpolar). At the equator, the whole celestial sphere can be seen in a year; the celestial poles lie on the horizons to the north and south, and all objects rise and set. At mid-latitudes, all of one celestial hemisphere is visible over a year, plus part of the other one. Only some objects are circumpolar.

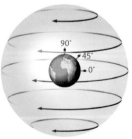

Apparent motion
Seen from various locations on the surface of the globe, celestial objects appear to move differently.

North Pole (90°N)
All objects move without rising or setting.

Mid-latitude (45°N)
Some objects rise and set; others are circumpolar.

Equator (0°)
All celestial objects rise and set.

TIME OF NIGHT AND TIME OF YEAR
As the Earth spins, the stars seem to move across the sky; hence our view of the sky changes as the night progresses. The celestial sphere turns once in the time it takes the Earth to spin on its axis relative to the stars – 23 hours 56 minutes. But the time between successive noons (a mean-time day) is longer, 24 hours, since the Earth is also orbiting the Sun. This means a star will rise four minutes earlier each night, as measured in mean time. The Earth's orbit of the Sun also means that a star near the celestial equator that is in the night sky (and hence visible) at one time of year will be in the daytime sky (and hence invisible) six months later.

Time of night
In this picture, taken with a long exposure, the images of the stars have been drawn into curving lines around the north celestial pole by the Earth's rotation.

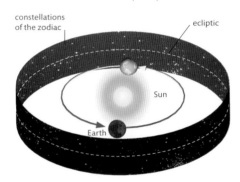

constellations of the zodiac
ecliptic
Sun
Earth

ZODIAC
During one year, the Sun passes in front of a band of constellations (see p.18) along the ecliptic. Collectively known as the zodiac, they form an area in which the planets are always to be seen. The dates when the Sun is in each constellation no longer accord with the dates attached to the astrological house of the same name.

The ecliptic and the zodiac
As the Earth moves around the Sun, the Sun appears to move against the background of stars.

CELESTIAL COORDINATES

Positions of objects on the celestial sphere are measured in coordinates called right ascension (RA) and declination (dec.). Right ascension is the equivalent of longitude on the Earth. It is measured in hours, anticlockwise around the celestial equator,

starting at the point where the Sun moves northwards across the equator each year (the vernal equinox). Declination, the equivalent of latitude on the Earth, is measured from 0 degrees at the celestial equator to 90 degrees at the celestial poles.

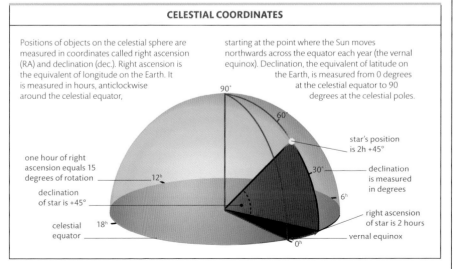

one hour of right ascension equals 15 degrees of rotation

declination of star is +45°

celestial equator

90°
60°
30°
12ʰ
18ʰ
6ʰ
0ʰ

star's position is 2h +45°
declination is measured in degrees
right ascension of star is 2 hours
vernal equinox

STAR NAMES AND CONVENTIONS

THE SKY is divided into 88 interlocking areas known as constellations, whose boundaries were defined by the International Astronomical Union in 1930. Within these areas, astronomers use an agreed system of names, letters, and numbers to identify celestial objects.

THE CONSTELLATIONS

In his book *The Almagest* (c. 150CE), the Greek astronomer Ptolemy listed 48 constellations, many of them depicting characters or creatures from Greek myth. Others were added later, making the modern total of 88. Many of the more recent constellations lie in the far southern sky that was below the horizon of the Greeks and depict scientific or technical instruments or exotic animals. On star charts, some of the stars in a constellation may be joined by lines to form a pattern that symbolizes the object after which the constellation is named. However, the resemblance to the real object is usually slight, and the way the linking lines are arranged varies between charts. Constellations are often referred to by a set of standard three-letter abbreviations – for example, Sagittarius is abbreviated to Sgr, Ursa Major to UMa.

star forming part of constellation pattern

linking line

CEN

LUP

CRU

NOR   CIR

MUS

ARA

TrA

APS   CHA   V

TEL

MEN

PAV   OCT   + south celestial pole

HYI

IND

TUC

Constellations around the south celestial pole

constellation boundary | abbreviated constellation name

MAGNITUDE

The brightness of an object in the night sky is called its apparent magnitude, and depends on both its actual luminosity and its distance from the Earth. Astronomers use a numerical scale to describe the magnitude of celestial objects. Bright stars are given low or negative numbers, while faint objects are given high numbers. Under good conditions, it is possible to see stars down to about magnitude 6 with the naked eye, while binoculars or a telescope are needed to see objects with higher values. The scale is logarithmic, so a star of magnitude 1 is 100 times brighter than a star of magnitude 6.

Apparent magnitude
Betelgeuse is about 10,000 times brighter than the Sun, yet it looks fainter than Sirius, which is only 20 times brighter than the Sun, because it is 50 times further away.

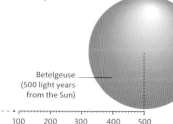

Sun

Sirius (8.6 light years from the Sun)

Betelgeuse (500 light years from the Sun)

0 1 2 3 4 5 6 7 8 9 10   100   200   300   400   500

STAR NAMES

There is no single system used to assign names to stars. Instead, several different schemes are used to name the various stars in a constellation. A large number of stars are referred to by letters or numbers, linked to the (Latin) genitive case of the name of the constellation; the brightest stars are given Greek letters, called Bayer letters, which are assigned alphabetically in roughly the order of brightness – hence the brightest star in Cygnus is named Alpha (α) Cygni. Most naked-eye stars have Flamsteed numbers (e.g. 61 Cygni), which run in order of right ascension. Some stars not in the Flamsteed catalogue are identified by upper- and lower-case letters, especially in southern constellations. A separate lettering system, starting with the capital letter R, is used for certain variable stars (e.g. T Cygni). Some bright stars also have common names derived from Latin, Greek, or Arabic – e.g. Deneb (Arabic). Deep-sky objects (such as nebulae, clusters, and galaxies) are given numbers preceded by the letters M, NGC, or IC, taken from the Messier, New General, or Index catalogues.

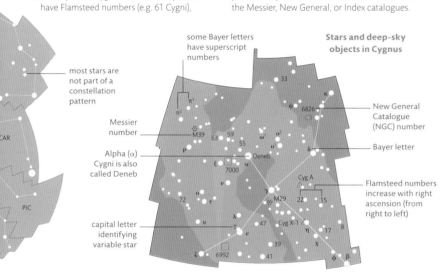

some Bayer letters have superscript numbers

Stars and deep-sky objects in Cygnus

most stars are not part of a constellation pattern

Messier number

Alpha (α) Cygni is also called Deneb

capital letter identifying variable star

New General Catalogue (NGC) number

Bayer letter

Flamsteed numbers increase with right ascension (from right to left)

DISTANCE

Astronomers measure distances to stars in the same way that a surveyor would: by taking bearings from two different places – in this case, from opposite sides of the Earth's orbit, six months apart. Even then, the shift in position, or parallax, is only tiny. If the shift is too small to detect, astronomers estimate how far away a star is by calculating how much light it emits and comparing this with how bright it appears in the sky.

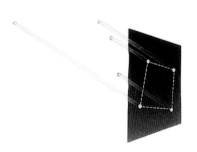

Patterns in three dimensions

The stars in the constellation patterns usually lie at various distances from the Earth and have no real connection with each other. For example, the stars in the Great Square of Pegasus range from 97 to nearly 400 light years away.

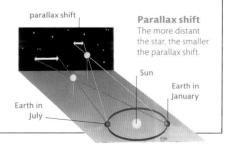

parallax shift

Parallax shift
The more distant the star, the smaller the parallax shift.

Sun

Earth in January

Earth in July

OBSERVING THE STARS

IDENTIFYING objects in the night sky for the first time can be a daunting experience. A good way to start is by learning to gauge distance and finding a few bright stars and key constellations.

Once these reference points have been identified, imaginary lines can be drawn outwards from them to find other stars and patterns, which in turn act as useful stepping-stones to the rest of the sky.

SCALE IN THE SKY

It can be difficult to judge how large a feature will appear in the sky by looking at a map. Fortunately, a hand held at arm's length acts as a convenient measuring scale. For example, an index finger easily covers the Moon or Sun, both of which are only half a degree across. The back of a hand is about 10 degrees across, the width of the bowl of the Plough (Big Dipper). To cover the Great Square of Pegasus (16 degrees across), the fingers must be splayed.

Using handspans
A hand held at arm's length can be used as a measuring device by both adults and children.

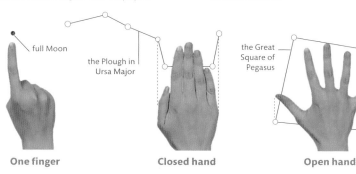

| One finger | Closed hand | Open hand |

NORTHERN SKIES

The Plough, or Big Dipper, which is high in northern-hemisphere skies in spring, is a key pattern in the northern sky. Two stars in the bowl of the Plough, Alpha and Beta Ursae Majoris, point to the north pole star, Polaris. On the other side of the celestial pole is the W-shaped constellation Cassiopeia. The Plough's bowl can also be used to locate the bright star Vega, which is prominent in northern skies in summer. A line extended from the curving handle of the Plough leads first to Arcturus, a beacon of spring skies, and then to Spica, the brightest star in Virgo. Also to the south of the bowl of the Plough are Leo and the stars Castor and Pollux in Gemini.

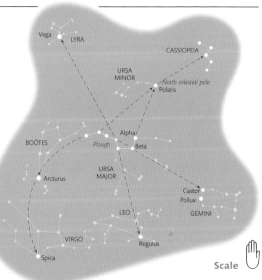

Northern signposts

Scale

MID-LATITUDE SKIES

Orion is a prominent constellation in the evening skies of northern winter and southern summer. The line of three stars that forms Orion's Belt points to the brightest star in the sky, Sirius (in Canis Major). Sirius marks one corner of a huge triangle (known in the northern hemisphere as the Winter Triangle). The other corners are marked by the bright stars Betelgeuse (in Orion) and Procyon (in Canis Minor). Another prominent star in Orion is Rigel. A line from Rigel through Betelgeuse points to Castor and Pollux in the neighbouring constellation of Gemini. On the other side of Orion is Aldebaran, the brightest star in Taurus; beyond Aldebaran, in the same direction, is the Pleiades open cluster. Almost directly to the north of Orion lies Capella, the brightest star in Auriga, which is overhead on January evenings from mid-northern latitudes.

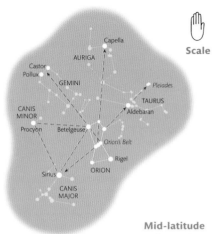

Scale

Mid-latitude signposts

SOUTHERN SKIES

Two bright stars in Centaurus and the familiar shape of Crux, the Southern Cross – all of which are at their highest on April and May evenings – are the starting points for finding a way around southern skies. A line from Alpha through nearby Beta Centauri points to Crux, the smallest constellation in the sky but also one of the most distinctive. The Southern Cross is not to be confused with the slightly larger False Cross, which is closer to the bright star Canopus. Unlike in the northern hemisphere, there is no bright star near the south celestial pole, but the longer axis of Crux points towards the pole, as does a line at right angles to the one linking Alpha and Beta Centauri (see also p.87). Canopus and Achernar form a large triangle with the south celestial pole.

Southern signposts

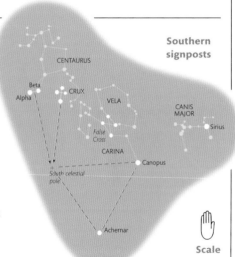

Scale

SKYWATCHING TIPS

- Always wear warm clothing – even summer evenings can be cold.
- Allow at least 10 minutes for your eyes to adapt to the dark before beginning serious observations.
- If needed, use only a dim red torch, so that your eyes remain adapted to the dark.

- Observe objects when they are well clear of the horizon wherever possible.
- To see faint objects more easily, look slightly to one side of them; their light will then fall on the more sensitive outer part of your eye. This technique is called averted vision.
- Keep a log of your observations, always noting the date and time.

Plastic-covered torch

BINOCULARS AND TELESCOPES

ASTRONOMERS use binoculars and telescopes to make objects appear brighter and larger. These instruments collect more light than the human eye, which means that much fainter objects can be seen than with the eye alone. They also magnify objects, so that small details can be distinguished. Binoculars are usually recommended for beginners; more serious study requires a telescope.

BINOCULARS OR A TELESCOPE?

For simple stargazing, binoculars are ideal, being portable and relatively cheap. They show many more stars than the naked eye, and their wide field of view makes them especially useful for looking at large areas of sky or objects such as comets. They also produce upright images. Telescopes have larger apertures and higher magnifications than binoculars, and show inverted images. The aperture of a telescope is fixed, but the magnification can be altered by changing the eyepiece. When assessing a telescope's performance the aperture is more important than the magnification (see below). Another critical requirement is a steady mounting, to prevent images from shaking. Most amateur astronomers use a telescope of "small" to "moderate" aperture – where small is less than 100mm (4in) and moderate is 100–250mm (4–10in); anything greater than this is considered large.

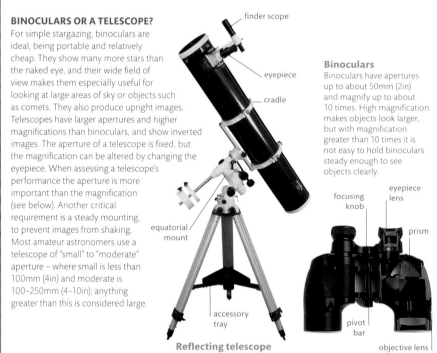

finder scope

eyepiece

cradle

equatorial mount

accessory tray

Reflecting telescope

Binoculars
Binoculars have apertures up to about 50mm (2in) and magnify up to about 10 times. High magnification makes objects look larger, but with magnification greater than 10 times it is not easy to hold binoculars steady enough to see objects clearly.

eyepiece lens

focusing knob

prism

pivot bar

objective lens

APERTURE AND MAGNIFICATION

Although magnification makes objects appear larger and closer, there are limits to the magnification that can be used. An image looks fainter as it gets bigger. As a guide, expect a maximum magnification of about twice the telescope aperture in millimetres (50 times the aperture in inches). No matter how large the aperture, as the magnification increases, the unsteadiness of the atmosphere, caused by air currents, becomes more noticeable, which limits the magnification that can be used. Astronomers call the steadiness of the atmosphere the "seeing".

Magnification
High powers of magnification are useful when viewing the Moon (shown here), planets, or close double stars. Low powers are better for diffuse objects such as comets, nebulae, and galaxies.

Naked eye   **Binoculars**   **Telescope**

REFRACTING TELESCOPES

Refractors have a main lens, called the object glass or objective lens, which collects and focuses light. The image is magnified by an eyepiece at the opposite end of the tube. In some small refractors, a prism is used to bend the light so that the eyepiece is more conveniently positioned for observation, as in the photograph here. Most small refractors have a simple type of mounting known as an altazimuth, which allows the telescope to tilt up and down and swivel from side to side.

Refractor
This small refracting telescope is on an altazimuth mounting.

How a refractor works

REFLECTING TELESCOPES

Most reflectors used by amateur observers are of the Newtonian type, in which light is collected by a concave primary mirror and deflected into an eyepiece at the side of the tube by a flat secondary mirror. Reflector mirrors have the advantage over refractor lenses in that they reflect all colours of light equally, so they give views free from false colour, which is a problem with budget refractors. The design shown here uses a smartphone to help find interesting objects.

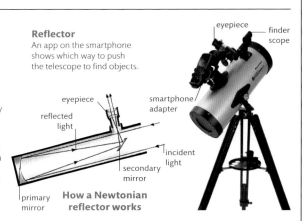

Reflector
An app on the smartphone shows which way to push the telescope to find objects.

How a Newtonian reflector works

CATADIOPTRIC TELESCOPES

Some telescopes, known as catadioptrics, combine mirrors with thin lenses known as corrector plates at the top of the tube. A popular design is the Schmidt-Cassegrain telescope or SCT. This has a secondary mirror mounted on the inside of the corrector plate to reflect light back through a hole in the centre of the primary mirror, where the eyepiece is inserted. Because the light path is folded back on itself within the tube, this design is more compact than the Newtonian.

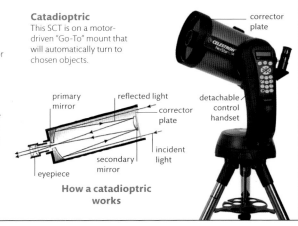

Catadioptric
This SCT is on a motor-driven "Go-To" mount that will automatically turn to chosen objects.

How a catadioptric works

ASTROPHOTOGRAPHY

MODERN DIGITAL cameras make it easy to take attractive pictures of the night sky, from wide-field views of the constellations to guided exposures that reveal objects too faint to be seen by eye.

With DSLR cameras, the normal camera lens can be replaced by a telescope, which acts like a very long telephoto lens, giving close-ups of deep-sky objects such as nebulae and galaxies.

DIRECT-CAMERA PHOTOGRAPHY

For astrophotography with a camera, select the manual setting to gain full control over the camera's aperture, length of exposure, and sensitivity. Set the lens to its widest aperture. The camera's sensitivity to light is described by an ISO number – the larger this number, the more sensitive it is, but also the grainier the image will be. An ISO 400 setting is a good compromise between speed and grain.

With the camera firmly supported, exposures of a second or two will show the brightest stars and planets. Longer exposures are possible if the camera is mounted on a tripod, but after ten seconds or so star images will start to trail (see p.17). To take sharp images of stars, the camera can be attached to a motor-driven mounting, which tracks the stars as the Earth rotates.

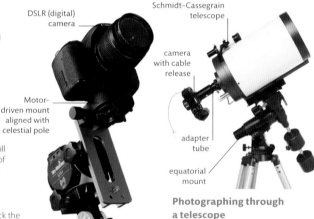

DSLR (digital) camera

Schmidt–Cassegrain telescope

camera with cable release

Motor-driven mount aligned with celestial pole

adapter tube

equatorial mount

Camera on tripod
A tripod-mounted camera will produce wide-angle views of the night sky with star trails.

Sky tracker
Portable driven mounts track the movement of the sky, allowing long-exposure photos of constellations and the larger deep-sky objects. To open the shutter for long periods a remote switch or cable release may be needed.

Photographing through a telescope
Adapters are available to allow a DSLR camera to replace a telescope eyepiece, using the telescope as a telephoto lens. If the telescope is motor-driven, long exposures of faint objects are possible.

ASTRO-CAMERAS

A wide variety of cameras are available for photography through a driven telescope. Some are designed for planetary imaging, and provide a rapid video stream. Specialized software is then used to select only the sharpest images and stack them. Others, usually with internal cooling to reduce electronic "noise", are suited to long exposures and give high-quality results. A laptop, tablet, or smartphone is needed to control the camera, store the images, and for image processing with photo editing software.

Cooled camera
Electronic cooling keeps the sensor below freezing so as to reduce noise.

Non-cooled camera
Planetary imaging requires a webcam-type video camera, where cooling is not needed. Many cooled cameras will perform both functions.

THE SOLAR SYSTEM

HOW THIS SECTION WORKS

THIS SECTION is about the Sun and the bodies in orbit around it. There are entries on the eight planets (including the Earth), each consisting of an illustrated description of the planet, advice on observation, and charts and diagrams that show where it can be located. Other entries cover the Sun itself, the Moon, and smaller bodies such as dwarf planets, comets, and meteors.

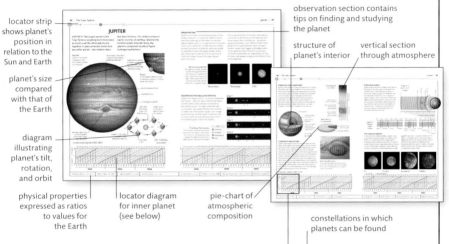

locator strip shows planet's position in relation to the Sun and Earth

planet's size compared with that of the Earth

diagram illustrating planet's tilt, rotation, and orbit

physical properties expressed as ratios to values for the Earth

locator diagram for inner planet (see below)

pie-chart of atmospheric composition

observation section contains tips on finding and studying the planet

structure of planet's interior

vertical section through atmosphere

Planet locators

The planets are always to be found in one of the 13 constellations of the zodiac (see p.17). The constellation in which the planets Mercury, Venus, Mars, Jupiter, and Saturn can be found at a particular time is shown on a strip diagram.

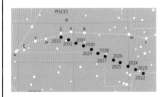

The outer planets

Uranus and Neptune seem to move relatively slowly. Their location in the night sky is therefore shown on more conventional charts.

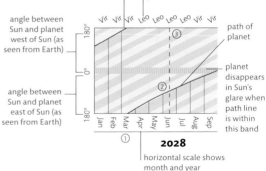

angle between Sun and planet west of Sun (as seen from Earth)

angle between Sun and planet east of Sun (as seen from Earth)

constellations in which planets can be found

path of planet

planet disappears in Sun's glare when path line is within this band

horizontal scale shows month and year

HOW TO USE THE STRIP DIAGRAM

1. Find the date on the horizontal scale. For example, to find Jupiter in June 2028 on the segment above, first locate June at the bottom of the diagram.

2. Read upward to the curved path line, whose position indicates the planet's separation from the Sun.

3. The constellation in which the planet appears in a particular month is listed along the top of the diagram – in June 2028, Jupiter is in Leo. Use the charts in the monthly sky guide section to find the location of Leo.

THE SUN

THE SUN is our local star. It is a ball of incandescent gas of average size, temperature, and brightness compared with other stars. The Sun dominates the Solar System: it is the source of heat and light for all the planets, and its great mass exerts a gravitational pull that keeps them in orbit around it. WARNING: Never look directly at the Sun through binoculars or a telescope.

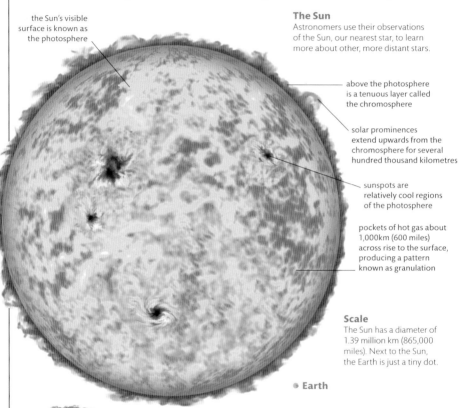

the Sun's visible surface is known as the photosphere

The Sun
Astronomers use their observations of the Sun, our nearest star, to learn more about other, more distant stars.

above the photosphere is a tenuous layer called the chromosphere

solar prominences extend upwards from the chromosphere for several hundred thousand kilometres

sunspots are relatively cool regions of the photosphere

pockets of hot gas about 1,000km (600 miles) across rise to the surface, producing a pattern known as granulation

Scale
The Sun has a diameter of 1.39 million km (865,000 miles). Next to the Sun, the Earth is just a tiny dot.

● **Earth**

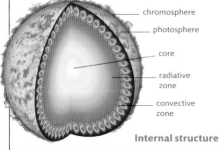

chromosphere

photosphere

core

radiative zone

convective zone

Internal structure

Structure
The Sun's energy is generated at its core, where temperatures are estimated to reach 15 million°C (27 million°F). Under these extreme conditions, hydrogen is converted to helium by nuclear fusion. Energy from these reactions travels outwards, first by radiation and then by convection, eventually reaching the photosphere. Above the photosphere is a less dense layer of gas, the chromosphere, from which bright clouds called prominences extend into the rarefied outermost region, the corona.

Diameter (Earth = 1): 109	Mass (Earth = 1): 333,000	Volume (Earth = 1): 1,300,000

SUNSPOTS

Sunspots, dark patches on the Sun's surface, are actually areas of cooler gas. They can last from a few days to several months, and range in size from a few hundred kilometres across to complex groups extending for 100,000km (60,000 miles). Occasional eruptions near sunspots, called flares, shoot out atomic particles that reach the Earth and can cause shortwave radio blackouts.

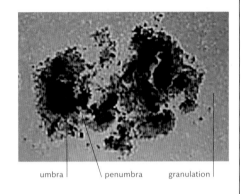

umbra   penumbra   granulation

SUNSPOT CYCLES

Sunspot numbers rise and fall in a cycle lasting about 11 years. At the start of the cycle, spots are few in number and appear away from the equator. Approaching solar maximum, the spots become more numerous (with up to 100 visible at one time) and appear closer to the equator.

Sunspot group

The centre of a sunspot, known as the umbra, is its darkest, coolest part, at about 4,500°C (8,100°F). The umbra is surrounded by a lighter, warmer area called the penumbra.

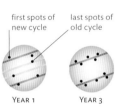

first spots of new cycle   last spots of old cycle

YEAR 1   **YEAR 3**

Solar minimum

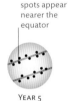

spots appear nearer the equator

YEAR 5

number of spots falls

YEAR 7

Solar maximum

last spots of old cycle   first spots of new cycle

YEAR 9   **YEAR 11**

Solar minimum

OBSERVING SUNSPOTS

Looking at the Sun with binoculars or a telescope can cause blindness. The only safe way to observe the Sun is to project its image on to a white surface, as shown below. The rim of the Sun appears darker than its centre, an effect known as limb darkening. Sunspots appear as small, dark marks that change position daily as the Sun rotates.

sunspot

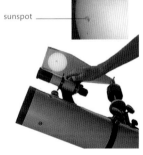

1. For safety, place a cap over the finder scope. If you are using a refractor, also cap the main lens.

2. Aim the telescope at the Sun by minimizing the size of the shadow cast by the telescope or the finder.

3. Remove the lens cap (if using a refractor), and focus the Sun's image on to a piece of white card.

| Density (water = 1): 1.41 | Gravity (Earth = 1): 27.94 | Surface temperature: 5,500°C (9,900°F) | Magnitude: -26.7 |

ECLIPSES OF THE SUN

Occasionally, the Moon passes in front of the Sun, briefly blocking its light to cause a solar eclipse. At least two solar eclipses occur each year, but not everyone will see them. During a solar eclipse, the Moon's shadow falls on the Earth. Anyone within the dark inner part of the shadow (the umbra) will see a total eclipse. From the lighter, outer shadow (the penumbra), a partial eclipse is seen. Because the umbra is quite narrow, total solar eclipses are rare at any one place on the Earth.

When the Sun's disc is entirely covered, its faint outer halo of gas, the corona, becomes visible. Scientists travel the world for the rare glimpses of the corona that an eclipse affords. A total eclipse can last for up to 7½ minutes, but 2 to 3 minutes is more usual; a partial eclipse can last for up to 3 to 4 hours. As with all solar observations, an eclipse is best viewed by projecting the Sun's image, but when the Sun is totally eclipsed the corona can safely be viewed with binoculars.

Sequence of a total solar eclipse

1. A partial eclipse – the Moon moves across the face of the Sun.

2. Just before totality, a shape similar to a diamond ring is seen.

3. At totality, the Sun's faint but extensive corona is visible.

4. The Moon moves away again and the eclipse ends.

How a solar eclipse occurs

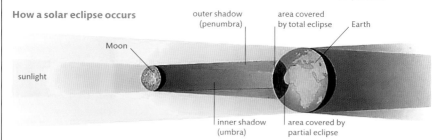

outer shadow (penumbra)

area covered by total eclipse

Earth

Moon

sunlight

inner shadow (umbra)

area covered by partial eclipse

ANNULAR ECLIPSE

When the Moon is at its furthest from the Earth, it does not appear quite big enough to cover the Sun completely. This gives rise to an annular eclipse, so named because a ring, or annulus, of light remains visible at mid-eclipse.

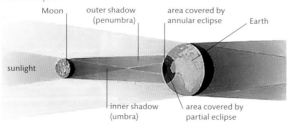

Moon

outer shadow (penumbra)

area covered by annular eclipse

Earth

sunlight

inner shadow (umbra)

area covered by partial eclipse

How an annular eclipse occurs

Ring of sunlight
An annulus of light is seen around the eclipsing Moon. The photosphere (rather than the corona) is visible here.

| Diameter (Earth = 1): 109 | Mass (Earth = 1): 333,000 | Volume (Earth = 1): 1,300,000 |

MERCURY

MERCURY is the closest planet to the Sun and is difficult to see because it is always low in the sky in the evening or morning twilight. It is a small, rocky planet, less than half the diameter of the Earth, and resembles our own Moon in appearance, being covered with craters. Mercury has no air or water. During the day, its surface is scorched by intense sunlight, while at night its temperature plummets far below freezing.

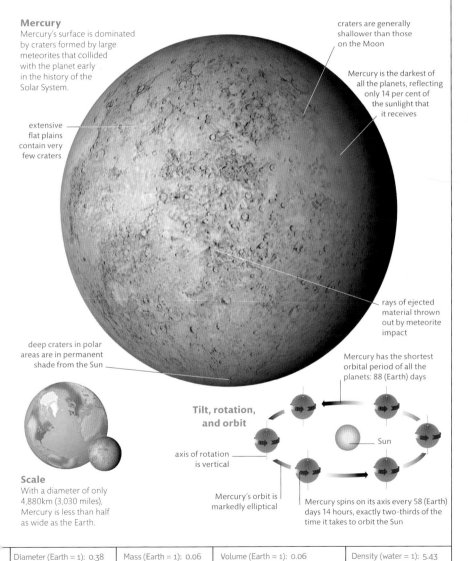

Mercury
Mercury's surface is dominated by craters formed by large meteorites that collided with the planet early in the history of the Solar System.

craters are generally shallower than those on the Moon

Mercury is the darkest of all the planets, reflecting only 14 per cent of the sunlight that it receives

extensive flat plains contain very few craters

rays of ejected material thrown out by meteorite impact

deep craters in polar areas are in permanent shade from the Sun

Mercury has the shortest orbital period of all the planets: 88 (Earth) days

Tilt, rotation, and orbit

axis of rotation is vertical

Sun

Scale
With a diameter of only 4,880km (3,030 miles), Mercury is less than half as wide as the Earth.

Mercury's orbit is markedly elliptical

Mercury spins on its axis every 58 (Earth) days 14 hours, exactly two-thirds of the time it takes to orbit the Sun

| Diameter (Earth = 1): 0.38 | Mass (Earth = 1): 0.06 | Volume (Earth = 1): 0.06 | Density (water = 1): 5.43 |

OBSERVATION

Of the five planets visible to the naked eye, Mercury is the most difficult to spot because it is always close to the Sun. Binoculars are usually needed to locate it. The best time to observe Mercury is around greatest elongation (see p.15), when it is furthest from the Sun. At that time, at a magnification of 250 times Mercury will appear as large as the full Moon looks to the naked eye. Like Venus and the Moon, Mercury's disc shows phases, which can be seen through a small telescope. Moderate to large apertures show dark, smudgy surface markings.

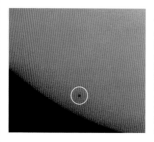

Transits

Occasionally, Mercury can be seen crossing the Sun's disc, an event known as a transit. In this photograph, taken at the transit of 2019, Mercury appears as a small, round dot. Future transits will occur on 13 November 2032, 7 November 2039, and 7 May 2049.

Mercury in the evening sky

At optimal viewing times, Mercury resembles a first-magnitude star. In this picture, it can be seen near the horizon, above and to the right of the large tree.

ATMOSPHERE AND CLIMATE

Mercury is airless and waterless. The barren, rocky surface is blasted by solar radiation, which raises its temperature to above 450°C (840°F) at noon on the equator when the planet is nearest the Sun. However, some shaded areas in craters near the poles remain permanently below the freezing point of water, and signs of ice have been detected there. At night the surface temperature plummets to below -180°C (-290°F).

Location of Mercury 2022–2033

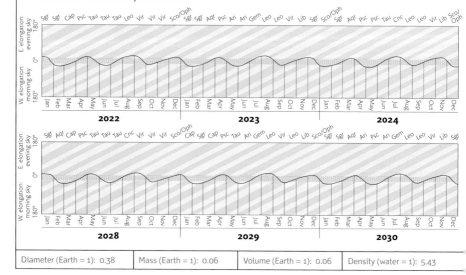

| Diameter (Earth = 1): 0.38 | Mass (Earth = 1): 0.06 | Volume (Earth = 1): 0.06 | Density (water = 1): 5.43 |

GEOGRAPHY AND GEOLOGY

Mercury looks much like our own Moon, since the rocky surfaces of both bodies have been extensively cratered by impacts from meteorites, and Mercury's diameter is only 40 per cent greater than that of the Moon. One distinct difference is that Mercury has cliffs up to 1.5km (1 mile) high running for several hundreds of kilometres, probably caused by shrinking and wrinkling of the planet as it cooled. Mercury is thought to have a large iron core that occupies three-quarters of its diameter.

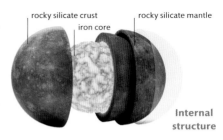

rocky silicate crust   rocky silicate mantle
iron core

Internal structure

CRATERS

As on the Moon, the youngest craters on Mercury are bright and surrounded by rays of material ejected by the force of the meteorite impact. Between Mercury's main craters are older, smoother areas called inter-crater plains, dotted with small craters. Secondary craters, caused by debris from the main impact, lie closer to the main craters than on the Moon because of Mercury's stronger gravity. Mercury has one large lowland plain, the Caloris Basin, which resembles the mare areas of the Moon. There are no known volcanic craters on Mercury.

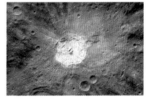

Kuiper

Rays of ejected material surround the crater Kuiper, 62km (39 miles) in diameter. This image was taken by the Messenger spacecraft in 2011.

The Caloris Basin

Part of this basin, 1,300km (800 miles) across, can be seen on the left of this picture. It was formed by a meteorite impact and later flooded with lava.

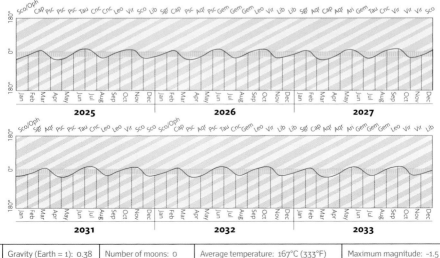

| Gravity (Earth = 1): 0.38 | Number of moons: 0 | Average temperature: 167°C (333°F) | Maximum magnitude: -1.5 |

VENUS

VENUS IS THE PLANET closest in size to the Earth, being only 650km (400 miles) smaller in diameter. It is the second planet from the Sun and comes within 40 million km (25 million miles) of the Earth, nearer than any other planet. It is the brightest object in the sky after the Sun and Moon, a result of both its closeness to us and the fact that it is entirely covered in bright clouds.

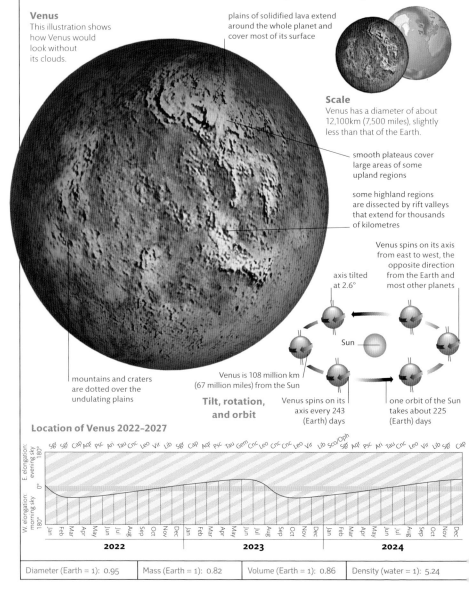

Venus
This illustration shows how Venus would look without its clouds.

plains of solidified lava extend around the whole planet and cover most of its surface

Scale
Venus has a diameter of about 12,100km (7,500 miles), slightly less than that of the Earth.

smooth plateaus cover large areas of some upland regions

some highland regions are dissected by rift valleys that extend for thousands of kilometres

Venus spins on its axis from east to west, the opposite direction from the Earth and most other planets

axis tilted at 2.6°

Sun

mountains and craters are dotted over the undulating plains

Venus is 108 million km (67 million miles) from the Sun

Tilt, rotation, and orbit

Venus spins on its axis every 243 (Earth) days

one orbit of the Sun takes about 225 (Earth) days

Location of Venus 2022–2027

E elongation: evening sky 180°

W elongation: morning sky 180°

0°

Sgr Sgr Cap Aqr Psc Ari Tau Cnc Leo Vir Lib Sgr Cap Aqr Psc Tau Gem Cnc Leo Cnc Cnc Leo Vir Lib Sco/Oph Sgr Aqr Psc Ari Tau Cnc Leo Vir Lib Sgr Cap

Jan Feb Mar Apr May Jun Jul Aug Sep Oct Nov Dec Jan Feb Mar Apr May Jun Jul Aug Sep Oct Nov Dec Jan Feb Mar Apr May Jun Jul Aug Sep Oct Nov Dec

2022 | **2023** | **2024**

| Diameter (Earth = 1): 0.95 | Mass (Earth = 1): 0.82 | Volume (Earth = 1): 0.86 | Density (water = 1): 5.24 |

OBSERVATION

Venus can be seen in the morning or evening sky, depending on whether it lies to the west or east of the Sun. It is always dazzling, reaching magnitude -4.7 at its brightest. During each orbit, Venus goes through a cycle of phases (see below). Binoculars will show the crescent phase, which occurs when Venus is closest to the Earth, but a small telescope is needed to see the full series. At half phase, a magnification of 75 times will show Venus the same size as the Moon appears to the naked eye. Due to its covering of clouds, the surface of the planet is invisible to us. Some dusky, V-shaped markings can be seen in the clouds, and the clouds over the poles can look brighter than the rest of the disc.

Naked-eye view
Venus is always close to the Sun in the sky, usually appearing near the horizon in the twilight. Its brilliance makes it easy to distinguish from adjacent stars.

Telescope view
Two crescent phases of Venus are shown here superimposed on one another.

PHASES

Venus completes a cycle of phases as it orbits the Sun. When on the far side of the Sun, and hence furthest from Earth, it appears as a full disc but is invisible behind the Sun's glare. As it moves closer to us, in the evening sky, it grows in apparent size. When at its greatest eastern elongation (see p.15), it passes through half phase to a crescent. The point when Venus is between the Sun and the Earth is known as inferior conjunction. It then moves into the morning sky, shrinking in apparent size but increasing in phase from a crescent via half to full again.

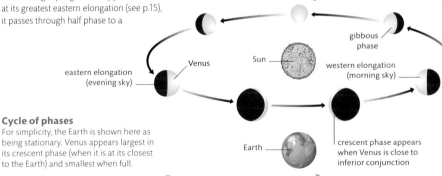

gibbous phase

Venus

Sun

western elongation (morning sky)

eastern elongation (evening sky)

Cycle of phases
For simplicity, the Earth is shown here as being stationary. Venus appears largest in its crescent phase (when it is at its closest to the Earth) and smallest when full.

Earth

crescent phase appears when Venus is close to inferior conjunction

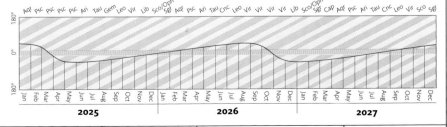

| Gravity (Earth = 1): 0.91 | Number of moons: 0 | Average temperature: 464°C (867°F) | Maximum magnitude: -4.7 |

ATMOSPHERE AND CLIMATE

Venus is enveloped in clouds of sulphuric acid, which lie 50-70km (30-45 miles) above its surface. Below the clouds, the atmosphere consists almost entirely of carbon-dioxide gas. At the surface, the atmospheric pressure is crushing, about 90 times greater than on the Earth. Temperatures soar to 464°C (867°F) due to the greenhouse effect (see below). Surface winds are gentle, only a few kilometres per hour, but above the clouds they reach hundreds of kilometres per hour. As a result, the clouds whip around the planet in four Earth days, far faster than the planet rotates on its axis, which takes 243 days. Like the planet itself, the clouds rotate from east to west.

Atmospheric composition

nitrogen (3.5%) and trace gases | carbon dioxide (96.5%)

Atmospheric structure

upper clouds of sulphuric acid

thin sulphuric acid haze

dense clouds producing rain of sulphuric acid

lower haze of dust and very small sulphuric acid droplets

lower atmosphere almost entirely carbon dioxide

THERMOSPHERE

TROPOSPHERE

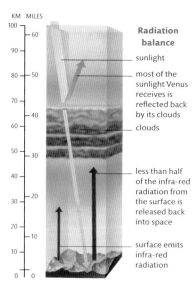

Radiation balance

sunlight

most of the sunlight Venus receives is reflected back by its clouds

clouds

less than half of the infra-red radiation from the surface is released back into space

surface emits infra-red radiation

THE GREENHOUSE EFFECT

Visible light from the Sun streams in through a planet's atmosphere, heating its surface. The surface then releases the heat as invisible infra-red radiation, but certain gases in the atmosphere, notably carbon dioxide, prevent some of the infra-red radiation from escaping directly into space, causing the temperature to increase. This is termed the greenhouse effect. In the case of the Earth, where carbon dioxide makes up only a small fraction of the atmosphere, the greenhouse effect raises the temperature by about 35°C (63°F). On Venus, where the atmosphere is almost entirely carbon dioxide, the increase in temperature due to the greenhouse effect is a massive 500°C (900°F), resulting in an average temperature of 464°C (867°F). Unlike on the Earth, the temperature on Venus differs by only a few degrees between the equator and the poles.

Location of Venus 2028-2033

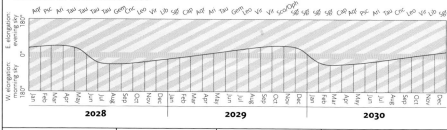

| Diameter (Earth = 1): 0.95 | Mass (Earth = 1): 0.82 | Volume (Earth = 1): 0.86 | Density (water = 1): 5.24 |

GEOGRAPHY AND GEOLOGY

Most of the surface of Venus is covered by undulating plains, which are dotted with mountains and craters formed by volcanic action and meteorite impacts. There are also two main upland regions, similar to continents on the Earth: Aphrodite Terra near the equator, which is over half the area of Africa; and Ishtar Terra in the far north, about the size of Australia. Ishtar contains the highest point on Venus, the Maxwell Montes mountain range, 11km (7 miles) high – taller than Mount Everest on the Earth. There are signs of recent lava flows, indicating that Venus may still be volcanically active today. Since Venus is so hot, all its water has evaporated.

silicate crust

iron and nickel core

rocky mantle

Internal structure

CRATERS

The surface of Venus is dotted with craters caused by meteorite impacts. Most are quite large, up to 300km (185 miles) in diameter, because large meteorites can penetrate the atmosphere whereas smaller ones burn up. Like the craters on the Moon, many on Venus have central peaks and are surrounded by ejected material. In addition, there are numerous craters formed by volcanic action.

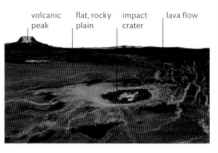

volcanic peak · flat, rocky plain · impact crater · lava flow

Maat Mons

This volcanic mountain in Aphrodite Terra is the second-highest peak on Venus.

summit is 8.5km (5.2 miles) high

Plains, mountains, and craters

Although its clouds prevent us from seeing its surface with a telescope, Venus has been mapped by radar. This image (and the two below) are based on data from the Magellan space probe.

Sapas Mons

This volcano has two central peaks surrounded by an expanse of solidified lava.

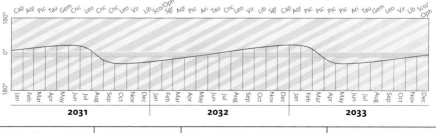

2031	2032	2033

| Gravity (Earth = 1): 0.91 | Number of moons: 0 | Average temperature: 464°C (867°F) | Maximum magnitude: -4.7 |

THE EARTH

THE EARTH, our home world, is the third planet from the Sun and the largest of the four rocky inner planets. Among the planets of the Solar System, the Earth is unique in having abundant surface water and an atmosphere rich in nitrogen and oxygen. It is also the only planet known with certainty to harbour life, supporting a complex population of plants and animals.

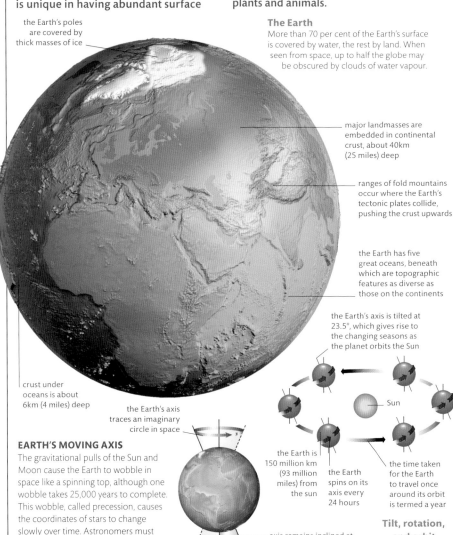

the Earth's poles are covered by thick masses of ice

The Earth
More than 70 per cent of the Earth's surface is covered by water, the rest by land. When seen from space, up to half the globe may be obscured by clouds of water vapour.

major landmasses are embedded in continental crust, about 40km (25 miles) deep

ranges of fold mountains occur where the Earth's tectonic plates collide, pushing the crust upwards

the Earth has five great oceans, beneath which are topographic features as diverse as those on the continents

the Earth's axis is tilted at 23.5°, which gives rise to the changing seasons as the planet orbits the Sun

crust under oceans is about 6km (4 miles) deep

the Earth's axis traces an imaginary circle in space

Sun

the Earth is 150 million km (93 million miles) from the sun

the Earth spins on its axis every 24 hours

the time taken for the Earth to travel once around its orbit is termed a year

Tilt, rotation, and orbit

EARTH'S MOVING AXIS

The gravitational pulls of the Sun and Moon cause the Earth to wobble in space like a spinning top, although one wobble takes 25,000 years to complete. This wobble, called precession, causes the coordinates of stars to change slowly over time. Astronomers must take this into account when measuring star positions.

23.5°

axis remains inclined at the same angle to the vertical as it wobbles

| Diameter: 12,756km (7,926 miles) | Mass: 5.98x10²¹ tonnes | Volume: 1.08x10¹² km³ (2.6 x10¹¹ mile³) |

ATMOSPHERE AND CLIMATE

The layer of gases that surrounds the Earth plays an important part in protecting and sustaining life. Not only does it shield the surface from excessive solar radiation and meteorite impacts, but its density, combined with Earth's favourable distance from the Sun, provides the right temperature for water to exist as a liquid. Water and carbon dioxide are used as food by plants, which in turn release the oxygen that is essential for animal life. Circulation in the atmosphere is also an important mechanism by which heat energy is redistributed polewards from the equator.

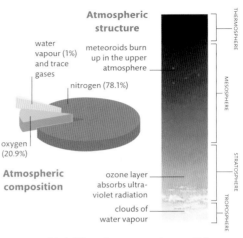

Atmospheric structure

water vapour (1%) and trace gases

meteoroids burn up in the upper atmosphere

nitrogen (78.1%)

oxygen (20.9%)

Atmospheric composition

ozone layer absorbs ultra-violet radiation

clouds of water vapour

THERMOSPHERE

MESOSPHERE

STRATOSPHERE

TROPOSPHERE

GEOGRAPHY AND GEOLOGY

The Earth's crust is made up of sections, called plates, that float on the surface of the viscous mantle and slowly move apart or slip beneath one another. Mountain ranges, volcanoes, and earthquakes occur along the margins of the plates. The Earth has a core of iron and nickel, the outer part of which is liquid. Motions in the liquid outer core give rise to the Earth's magnetic field, which extends into space, creating a magnetic cocoon around our planet termed the magnetosphere.

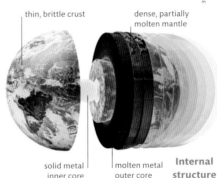

thin, brittle crust

dense, partially molten mantle

solid metal inner core

molten metal outer core

Internal structure

LIGHT POLLUTION

Light pollution is caused by artificial lights shining into the sky. The problem is greatest in urban areas and is so extensive that in some developed countries there are now few places with truly dark skies. As well as being a waste of energy, light pollution is a growing nuisance to astronomers, as relatively faint celestial objects become lost in the glare from surface lighting.

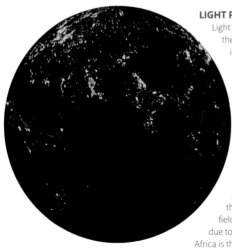

The Earth at night

At the top left of this picture are the lights of industrialized Europe, and at the top right is Japan. The yellow areas, such as those around the Middle East, represent gas burned off in oil fields, while the red specks in equatorial Africa are due to agricultural burning. The line of light in North Africa is the Nile Valley.

| Density (water = 1): 5.52 | Gravity: 1 | Number of moons: 1 | Average temperature: 15°C (59°F) |

THE MOON

THE MOON is the Earth's only natural satellite. It is so close to the Earth – at an average distance of only 384,400km (238,900 miles) – that even binoculars will reveal features on its surface in fascinating detail. The Moon has no air or liquid water, and hence no weather; it is also lifeless and without geological activity. Its main landforms have all resulted from meteorite impact.

THE NEAR SIDE OF THE MOON

The Moon's distinctive lowland plains, the maria, are labelled here; other features are cross-referred to the panel on the facing page.

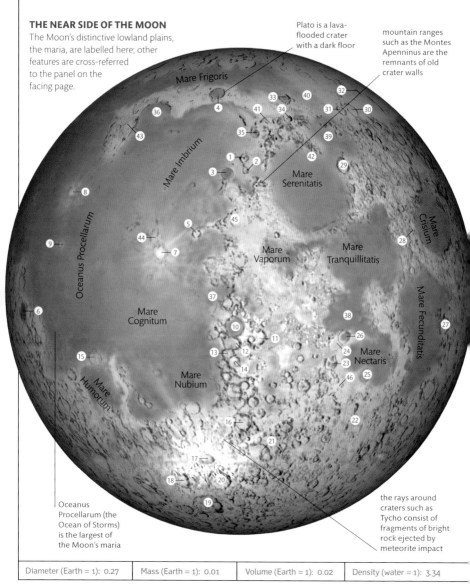

Plato is a lava-flooded crater with a dark floor

mountain ranges such as the Montes Apenninus are the remnants of old crater walls

Mare Frigoris

Mare Imbrium

Mare Serenitatis

Mare Crisium

Oceanus Procellarum

Mare Vaporum

Mare Tranquillitatis

Mare Fecunditatis

Mare Cognitum

Mare Nectaris

Mare Humorum

Mare Nubium

Oceanus Procellarum (the Ocean of Storms) is the largest of the Moon's maria

the rays around craters such as Tycho consist of fragments of bright rock ejected by meteorite impact

| Diameter (Earth = 1): 0.27 | Mass (Earth = 1): 0.01 | Volume (Earth = 1): 0.02 | Density (water = 1): 3.34 |

Scale
The Moon's diameter is 3,476km (2,160 miles), just over a quarter of that of the Earth.

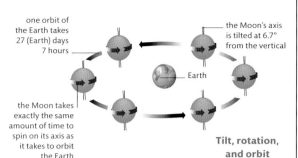

one orbit of the Earth takes 27 (Earth) days 7 hours

the Moon's axis is tilted at 6.7° from the vertical

Earth

the Moon takes exactly the same amount of time to spin on its axis as it takes to orbit the Earth

Tilt, rotation, and orbit

OBSERVING THE MOON

Since the Moon spins on its axis exactly once in the time it takes to orbit the Earth, the same hemisphere, termed the near side, is always turned towards Earth. Hence, we only ever see about half of the lunar surface. The main features can be seen with the naked eye. Dark lowland plains, known as maria, contrast with the bright highlands. Binoculars or a small telescope reveal innumerable craters and mountain ranges. Surface relief is most conspicuous near the border between lit and unlit areas, called the terminator, where shadows are longest. When the surface is illuminated at high angles near full Moon, it appears washed out, and even large craters can vanish. However, some features need high contrast to be seen and are at their most prominent near full Moon; these include the dark maria and the bright rays around some craters.

imaginary point always faces the Earth

Day 14

Moon

Day 7

Day 21

Earth

Day 1

Captured rotation
The Earth's gravitational influence has slowed the Moon's rotation so that it is exactly synchronized with its orbit.

The far side
The Moon's far side, invisible from the Earth, has been photographed from space.

KEY TO LUNAR FEATURES

Craters

1. Aristillus	13. Alpetragius	25. Fracastorius	**Lowland features**
2. Autolycus	14. Arzachel	26. Theophilus	36. Sinus Iridum
3. Archimedes	15. Gassendi	27. Langrenus	37. Sinus Medii
4. Plato	16. Walter	28. Proclus	38. Sinus Asperitatis
5. Eratosthenes	17. Tycho	29. Posidonius	39. Lacus Somniorum
6. Grimaldi	18. Longomontanus	30. Atlas	40. Lacus Mortis
7. Copernicus	19. Clavius	31. Hercules	**Upland features**
8. Aristarchus	20. Maginus	32. Endymion	41. Montes Alpes
9. Marius	21. Stöfler	33. Aristoteles	42. Montes Caucasus
10. Ptolemaeus	22. Piccolomini	34. Eudoxus	43. Montes Jura
11. Albategnius	23. Catharina	35. Cassini	44. Montes Carpatus
12. Alphonsus	24. Cyrillus		45. Montes Apenninus
			46. Rupes Altai

Gravity (Earth = 1): 0.17	Average temperature: -18°C (0°F)	Maximum magnitude: -12.7

PHASES OF THE MOON

The Moon's phases arise because we see differing amounts of its sunlit side as it orbits the Earth. At new Moon, the near side of the Moon is in complete shadow. As it orbits the Earth, the visible area grows, or waxes, from a crescent via a gibbous phase to full, when the disc is fully illuminated. The phases then occur in reverse, as the visible area shrinks, or wanes. When the Moon is a thin crescent, the unlit part can be seen faintly illuminated by light reflected from the Earth, an effect known as earthshine.

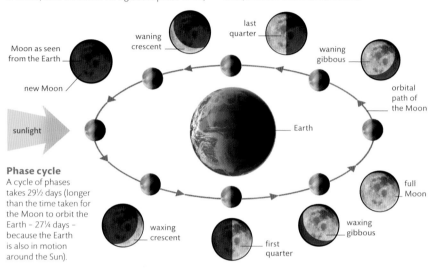

Moon as seen from the Earth

new Moon

waning crescent

last quarter

waning gibbous

orbital path of the Moon

sunlight

Earth

full Moon

Phase cycle

A cycle of phases takes 29½ days (longer than the time taken for the Moon to orbit the Earth – 27¼ days – because the Earth is also in motion around the Sun).

waxing crescent

first quarter

waxing gibbous

ECLIPSES OF THE MOON

Occasionally, the full Moon moves into the Earth's shadow and is eclipsed. From the Moon's first entry into the shadow until it has completely re-emerged can take up to four hours, and the period during which the eclipse is total can be well over an hour. Up to three lunar eclipses may occur in a year, although in some years there are none. Lunar eclipses are widely visible on the Earth's surface, from wherever the Moon is above the horizon.

Sequence of a lunar eclipse

Sunlight bent into the Earth's shadow by its own atmosphere often prevents the eclipsed Moon disappearing completely and makes it appear red.

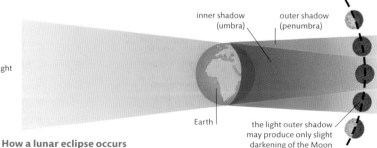

inner shadow (umbra)

outer shadow (penumbra)

sunlight

Earth

the light outer shadow may produce only slight darkening of the Moon

How a lunar eclipse occurs

| Diameter (Earth = 1): 0.27 | Mass (Earth = 1): 0.01 | Volume (Earth = 1): 0.02 | Density (water = 1): 3.34 |

HISTORY AND GEOLOGY

The Moon is thought to have formed about 4.5 billion years ago when a body the size of Mars struck the Earth, which itself was still less than 100 million years old. Debris from the collision was thrown into orbit and coalesced to form the Moon. The young Moon was bombarded by meteorites until about 3.9 billion years ago, when the storm of impacts subsided. Molten lava then oozed out from the interior, creating the maria, a process that continued for more than 2 billion years. Maria are scarce on the far side because the crust is thicker there, and so less lava reached the surface.

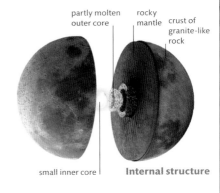

partly molten outer core — rocky mantle — crust of granite-like rock

small inner core

Internal structure

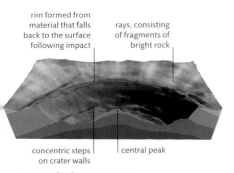

rim formed from material that falls back to the surface following impact

rays, consisting of fragments of bright rock

concentric steps on crater walls | central peak

Meteorite impact crater
Large craters often have a flat floor with one or more mountains at their centre. Since there is no erosion on the Moon, craters are well preserved from the time of impact.

LANDFORMS

The Moon's lowland maria bear few large craters. In contrast, the highlands, which are older, are saturated with craters of all sizes, from tiny pits to huge walled plains over 200km (125 miles) across. With the exception of a few small volcanic craters, they were all formed by meteorite impacts. Large craters often have central peaks, caused by rebounding of the crater floor during impact, and stepped walls caused by subsequent slumping. Young craters are bright and surrounded by rays of pulverized rock thrown out by the impact. The Moon also has valleys, termed rilles, of which there are two main types: those caused by faulting, which can be straight or curved; and sinuous rilles, which look like rivers but are actually lava channels.

Craters and rilles
Aristarchus (on the left) is a young ray crater. A long rille leads away from the adjacent crater Herodotus.

Overlapping craters
The crater Theophilus (on the right) has a central peak and terraced wall.

| Gravity (Earth = 1): 0.17 | Average temperature: –18°C (0°F) | Maximum magnitude: –12.7 |

MARS

MARS IS A small, rocky planet that in some ways is similar to the Earth. It has a 24-hour day, a pattern of seasons that resembles our own, and polar ice-caps. However, there are important differences: temperatures on Mars rarely rise above freezing, and the atmosphere is thin and contains almost no oxygen. Mars is often called the Red Planet because it is covered by red deserts.

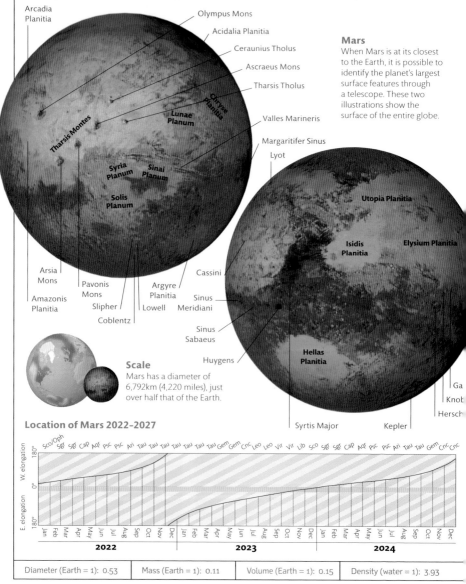

Arcadia Planitia
Olympus Mons
Acidalia Planitia
Ceraunius Tholus
Ascraeus Mons
Tharsis Tholus
Chryse Planitia
Lunae Planum
Valles Marineris
Margaritifer Sinus
Lyot
Tharsis Montes
Syria Planum
Sinai Planum
Solis Planum
Utopia Planitia
Isidis Planitia
Elysium Planitia
Arsia Mons
Cassini
Amazonis Planitia
Pavonis Mons
Argyre Planitia
Sinus Meridiani
Slipher
Lowell
Coblentz
Sinus Sabaeus
Hellas Planitia
Huygens
Syrtis Major
Kepler
Ga
Knob
Hersch

Mars
When Mars is at its closest to the Earth, it is possible to identify the planet's largest surface features through a telescope. These two illustrations show the surface of the entire globe.

Scale
Mars has a diameter of 6,792km (4,220 miles), just over half that of the Earth.

Location of Mars 2022–2027

| Diameter (Earth = 1): 0.53 | Mass (Earth = 1): 0.11 | Volume (Earth = 1): 0.15 | Density (water = 1): 3.93 |

OBSERVATION

Mars is easy to spot with the naked eye at or near times of opposition (see below), when it resembles a bright and distinctly orange-coloured star. Binoculars show only a tiny disc, but with a small telescope the planet's polar caps and the largest of its dark surface markings may be seen. For serious observing, a telescope with an aperture of 200mm (8in) or more is recommended. At its average opposition distance, a magnification of about 100 times makes Mars appear as large as the full Moon does to the naked eye.

Naked eye
Mars is the orange dot above the right-hand side of the spire.

Binoculars
Mars is the disc below centre in this picture.

Telescope
Large-scale surface features appear as light and dark markings.

Tilt, rotation, and orbit

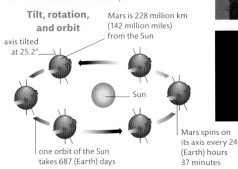

Mars is 228 million km (142 million miles) from the Sun

axis tilted at 25.2°

Sun

Mars spins on its axis every 24 (Earth) hours 37 minutes

one orbit of the Sun takes 687 (Earth) days

OPPOSITIONS

Mars comes to opposition (see p.15) roughly every two years and two months, but its highly elliptical orbit means that its distance from us at opposition varies appreciably. At close oppositions – every 15 or 17 years – it appears largest and brightest in the sky.

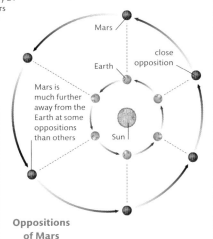

Mars

close opposition

Earth

Mars is much further away from the Earth at some oppositions than others

Sun

Oppositions of Mars

DATES OF OPPOSITIONS

DATE	WHERE MARS APPEARS	MAGNITUDE	DISTANCE FROM EARTH (MILLION KM)
8 December 2022	Taurus	-1.9	82
16 January 2025	Gemini	-1.4	96
19 February 2027	Leo	-1.2	101
25 March 2029	Virgo	-1.3	97
4 May 2031	Libra	-1.8	84
28 June 2033	Sagittarius	-2.5	64

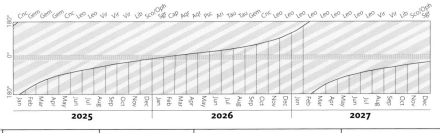

Cnc Gem Gem Gem Cnc Leo Leo Vir Vir Vir Lib Sco/Oph Sgr Cap Aqr Aqr Psc Ari Tau Tau Gem Cnc Leo Leo Leo Leo Leo Leo Leo Leo Vir Vir Lib Sco/Oph Sgr

180°
0°
180°

Jan Feb Mar Apr May Jun Jul Aug Sep Oct Nov Dec | Jan Feb Mar Apr May Jun Jul Aug Sep Oct Nov Dec | Jan Feb Mar Apr May Jun Jul Aug Sep Oct Nov Dec

2025 **2026** **2027**

| Gravity (Earth = 1): 0.38 | Number of moons: 2 | Average temperature: -63°C (-81°F) | Maximum magnitude: -2.8 |

ATMOSPHERE AND CLIMATE

Mars has an atmosphere that consists mostly of carbon dioxide. It is so thin that the pressure at the surface is less than 1 per cent of that on Earth, and the temperature is usually well below freezing. In winter, the atmosphere freezes at the poles, adding a layer of carbon-dioxide frost to the permanent polar caps of frozen water. Morning mists and high-altitude clouds, particularly in polar areas, can be seen from the Earth. Localized dust storms occur throughout the Martian year. However, when Mars is at its closest to the Sun – and temperature and wind speed are at their maximum – storms can cover the whole planet, enveloping it in dust.

Atmospheric structure

thin clouds of frozen carbon dioxide

isolated clouds and fog of icy water vapour

iron-rich red dust

THERMOSPHERE

STRATOSPHERE

TROPOSPHERE

Atmospheric composition

oxygen and carbon monoxide (0.2%), and trace gases

argon (1.9%)

nitrogen (2.6%)

carbon dioxide (95.1%)

MOONS

Mars has two small moons called Phobos and Deimos, with average diameters of about 22km (14 miles) and 12km (7 miles). Both moons are irregularly shaped and are thought to be asteroids that have been pulled into orbit around Mars by the planet's gravity. Phobos orbits Mars more quickly than Mars rotates on its axis, so from the planet's surface it appears to rise in the west and set in the east. Phobos and Deimos are too faint to be seen without a large telescope. Close-up photographs from space probes have shown that both are heavily cratered.

Stickney

Phobos
The largest crater on Phobos, called Stickney, is about 10km (6 miles) in diameter.

Deimos
The smaller of the two moons, Deimos has no large craters.

Phobos

Deimos

Moon orbits
Phobos and Deimos are shown here orbiting Mars (not to scale).

1   2   3   4   5   6   7   8   Scale in radiuses of Mars

Location of Mars 2028–2033

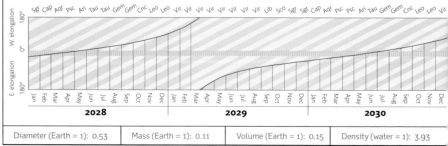

| Diameter (Earth = 1): 0.53 | Mass (Earth = 1): 0.11 | Volume (Earth = 1): 0.15 | Density (water = 1): 3.93 |

GEOGRAPHY AND GEOLOGY

Mars is a planet of two halves: the southern hemisphere is mainly upland, heavily cratered by meteorites; the northern hemisphere is smoother and on average a few kilometres lower in altitude. There are several huge volcanoes, the most impressive being Olympus Mons, a mountain 600km (400 miles) wide and 21km (13 miles) high, and often covered by white clouds that can be seen with a telescope on Earth. Another spectacular feature is a system of canyons, Valles Marineris, more than 4,000km (2,500 miles) long, one-fifth of the planet's circumference and visible through a telescope as a dark streak. The surface of Mars resembles a rock-strewn desert, coloured a rusty red by iron oxide and frozen to a depth of several kilometres. There are signs that water has flowed over the surface in the past, when the climate was warmer.

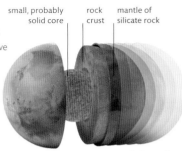

small, probably solid core   rock crust   mantle of silicate rock

Internal structure

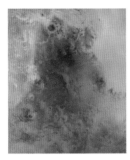

Syrtis Major
Once thought to be an area of vegetation, this elongated dark marking actually consists of darker rock and dust.

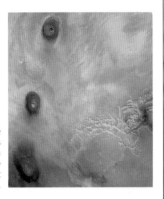

The Tharsis Bulge
There are three volcanoes in this upland region, forming a chain 2,000km (1,250 miles) long. The edge of the Valles Marineris is at the right of the picture.

Surface
Layers of sedimentary rock imaged by NASA's Curiosity rover are evidence that water flowed on the surface of Mars in the past.

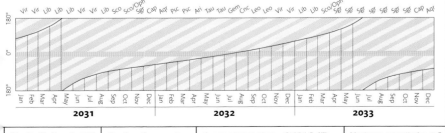

| Gravity (Earth = 1): 0.38 | Number of moons: 2 | Average temperature: -63°C (-81°F) | Maximum magnitude: -2.8 |

JUPITER

JUPITER IS THE largest planet in the Solar System, weighing more than twice as much as all the other planets put together. It spins on its axis faster than any other planet – one rotation takes less than 10 hours. The visible surface of Jupiter consists of swirling, distinctively banded clouds. Beneath them, the planet is composed mostly of liquid hydrogen and helium.

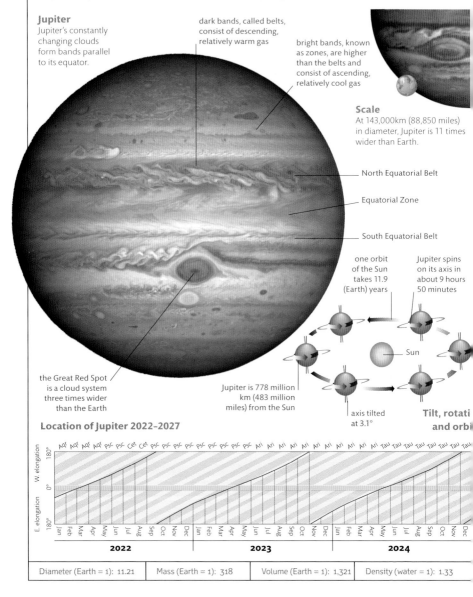

Jupiter
Jupiter's constantly changing clouds form bands parallel to its equator.

dark bands, called belts, consist of descending, relatively warm gas

bright bands, known as zones, are higher than the belts and consist of ascending, relatively cool gas

Scale
At 143,000km (88,850 miles) in diameter, Jupiter is 11 times wider than Earth.

North Equatorial Belt

Equatorial Zone

South Equatorial Belt

one orbit of the Sun takes 11.9 (Earth) years

Jupiter spins on its axis in about 9 hours 50 minutes

Sun

the Great Red Spot is a cloud system three times wider than the Earth

Jupiter is 778 million km (483 million miles) from the Sun

axis tilted at 3.1°

Tilt, rotati and orbi

Location of Jupiter 2022–2027

| Diameter (Earth = 1): 11.21 | Mass (Earth = 1): 318 | Volume (Earth = 1): 1,321 | Density (water = 1): 1.33 |

OBSERVATION

Jupiter is one of the most interesting planets for the amateur observer. It is usually the brightest planet after Venus (although at times Mars can become slightly brighter), and it is so large that even binoculars show it as a rounded disc. A small telescope reveals the most prominent cloud bands, which are usually the dark North and South Equatorial Belts, either side of the bright Equatorial Zone. An aperture of 75mm (3in) reveals some of the larger features within the clouds.

Over a period of about 10 minutes, these features can be seen to be moving. A small telescope also shows Jupiter's disc to be elliptical – the planet's equatorial diameter is over 9,000km (5,600 miles) greater than its polar diameter, due to its rapid rotation. Jupiter looks biggest and brightest when it is at opposition (see p.15), which occurs every 13 months. At that time, at a magnification of only 40 times it will appear as large as the full Moon looks to the naked eye.

Observing Jupiter
Through binoculars, Jupiter can be distinguished from stars because it appears as a disc rather than a point of light. Telescopes and CCD cameras show the surface of the disc in greater detail.

Binoculars

Telescope

CCD

OBSERVING THE GALILEAN MOONS

Jupiter's four largest moons – Io, Europa, Ganymede, and Callisto – were first seen in 1610 by the Italian scientist Galileo Galilei, and are now known as the Galilean moons. Through a small telescope or binoculars, they appear like faint stars lined up on either side of the planet's equator, and change position as they orbit the planet. Sometimes one or more may be missing, either passing in front of or behind Jupiter, or lost in its shadow.

Tracking the moons
The moons' positions are shown here over four nights. The moons are shown as coloured discs but appear through binoculars or a telescope as small points of light.

KEY
● Callisto (C)
● Europa (E)
● Ganymede (G)
● Io (I)

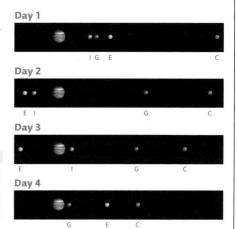

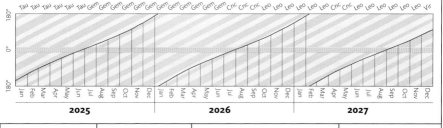

| Gravity (Earth = 1): 2.36 | Number of moons: c.80 | Average temperature: -144°C (-227°F) | Maximum magnitude: -2.9 |

STRUCTURE AND ATMOSPHERE

Jupiter is enveloped in a cloudy atmosphere about 1,000km (600 miles) deep. Clouds of different composition form at various levels depending on the temperature and pressure, which both increase with depth. Beneath the clouds there is no solid surface. Instead, hydrogen and helium are compressed into liquids by Jupiter's great gravitational pull. Deep in the interior, where liquid hydrogen acts like molten metal, convection produces a strong magnetic field that extends for millions of kilometres into space. At Jupiter's centre there is thought to be a rocky core.

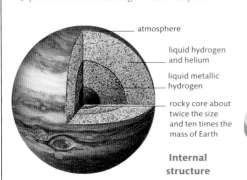

atmosphere

liquid hydrogen and helium

liquid metallic hydrogen

rocky core about twice the size and ten times the mass of Earth

Internal structure

Atmospheric structure

STRATOSPHERE

TROPOSPHERE

white clouds of ammonia crystals

dark orange clouds of ammonium hydrosulphide

bluish clouds of water ice and water droplets

Atmospheric composition

helium (10.2%) and trace gases

hydrogen (89.8%)

CLOUD BELTS AND STORMS

Jupiter's rapid rotation draws its clouds into light and dark bands. In the lighter areas, termed zones, gas rises from the warm interior and condenses to form high-altitude clouds. The darker clouds, called belts, occur at lower altitudes, where gas descends. The colours of the belts range from red and brown to blue, depending on the compounds that they contain. Individual cloud features seldom last more than a few weeks, but some white oval clouds have lasted for over half a century. The most prominent feature is the Great Red Spot, on the southern edge of the South Equatorial Belt. This swirling, high-altitude storm cloud rotates anti-clockwise once a week or so. It has been tracked since 1831 although a similar feature was seen in the 17th century.

The Great Red Spot
This image was taken from the Juno space probe. The colouring of the spot is thought to be due to red phosphorus, or perhaps carbon compounds.

Location of Jupiter 2028–2033

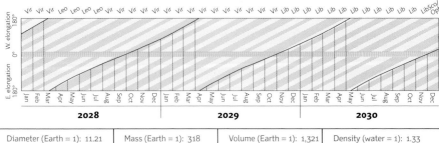

| Diameter (Earth = 1): 11.21 | Mass (Earth = 1): 318 | Volume (Earth = 1): 1,321 | Density (water = 1): 1.33 |

RINGS AND MOONS

Jupiter has about 80 known satellites, which fall into three groups: the inner eight, including the four Galilean moons, which have circular orbits in the planet's equatorial plane; the middle ten, with elliptical orbits inclined from 27 to 51 degrees to Jupiter's equator; and an outer swarm, which have elliptical and retrograde (east-to-west) orbits. The last two groups are probably captured asteroids. Apart from the Galilean moons, the satellites are small and faint – a large telescope is needed to see them. Jupiter has a faint ring of dust, over 100,000km (60,000 miles) wide.

Rings and innermost moons

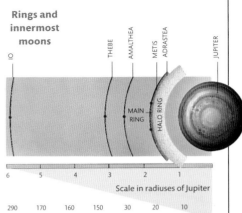

Scale in radiuses of Jupiter

Jupiter's largest outer moons

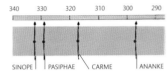

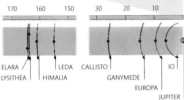

THE GALILEAN MOONS

Io, the innermost Galilean satellite, is similar in size to our Moon, having a diameter of 3,643km (2,264 miles). It is covered with yellow-orange sulphur, erupted by volcanoes that are still active. Its interior is kept molten by Jupiter's gravitational forces. Europa, 3,122km (1,940 miles) in diameter, is the smallest of the four. Its smooth, icy surface resembles a cracked eggshell. Ganymede – diameter 5,262km (3,270 miles) – is the largest moon in the Solar System, bigger than Mercury. Its complex, partly cratered surface has dark patches and lighter grooves. Outermost of the four moons is Callisto – diameter 4,821km (2,995 miles) – the dark surface of which is covered with impact craters.

Io   Europa   Ganymede   Callisto

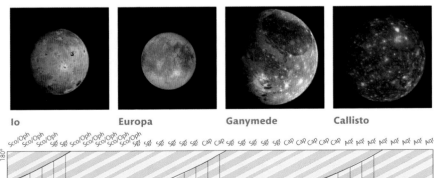

| Gravity (Earth = 1): 2.36 | Number of moons: c.80 | Average temperature: –144°C (–227°F) | Maximum magnitude: –2.9 |

SATURN

SATURN, THE MOST DISTANT planet known to ancient astronomers, is the second largest in the Solar System, and the easiest one to recognize through a telescope because of the broad, bright rings that encircle its equator. Like Jupiter, it has a cloudy atmosphere overlying an interior of liquid hydrogen and helium. Saturn has more than 80 moons, some orbiting among the rings.

Saturn
Saturn's best-known feature is its magnificent ring system.

Scale
Saturn is 120,500 km (75,000 miles) across, nine times wider than the Earth.

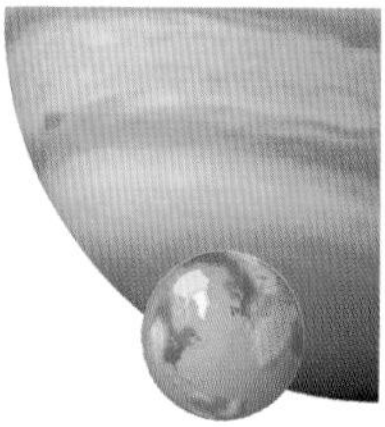

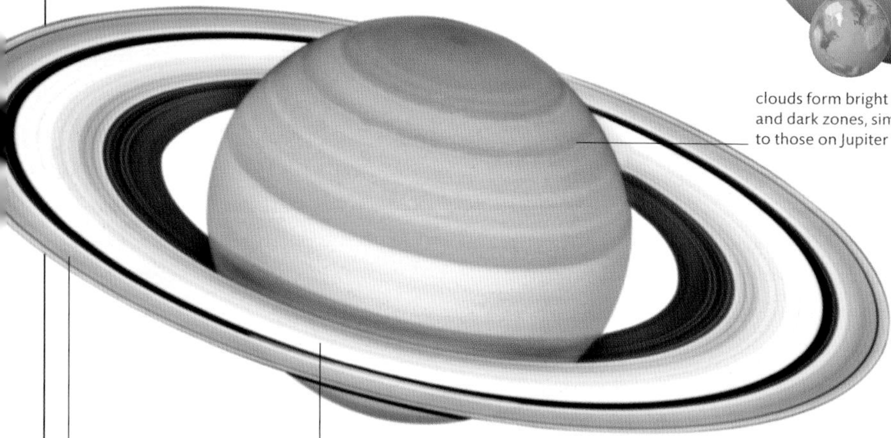

clouds form bright belts and dark zones, similar to those on Jupiter
in relation to their overall width, Saturn's rings are thinner than a sheet of paper
rings consist of icy lumps, ranging from tiny particles to pieces a few metres across

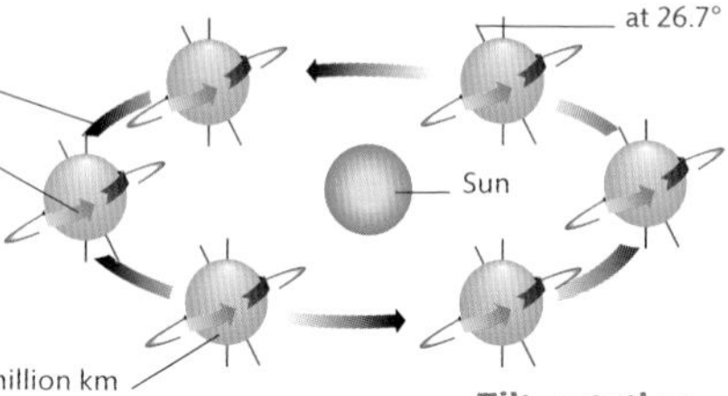

axis tilted at 26.7°
one orbit takes 29.5 (Earth) years
Saturn spins on its axis every 10 hours 14 minutes at the equator; the speed gradually decreases towards the poles
Sun
Saturn is 1,427 million km (886 million miles) from the Sun

Tilt, rotation, and orbit

Location of Saturn 2022–2027

Cap Cap Cap Cap Cap Cap Cap Cap Cap Cap Cap Cap Cap Cap Aqr Aqr Aqr Aqr Aqr Aqr Aqr Aqr Aqr Aqr Aqr Aqr Aqr Aqr Aqr Aqr Aqr Aqr Aqr Aqr Aqr Aqr
W. elongation
E. elongation
180°
0°
180°
Jan Feb Mar Apr May Jun Jul Aug Sep Oct Nov Dec
Jan Feb Mar Apr May Jun Jul Aug Sep Oct Nov Dec
Jan Feb Mar Apr May Jun Jul Aug Sep Oct Nov Dec
2022
2023
2024

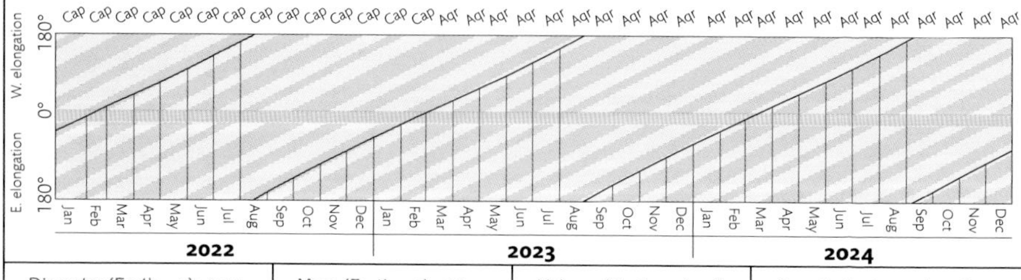

Diameter (Earth = 1): 9.14	Mass (Earth = 1): 95	Volume (Earth = 1): 764	Density (water = 1): 0.69

OBSERVATION
To the naked eye, Saturn looks like a bright, yellowish star. Binoculars just reveal a small disc, with a hint of elongation caused by its rings With a small telescope, the rings, which are over twice as wide as the planet itself, can be seen clearly. Of the five planets visible to the naked eye, Saturn is the slowest-moving, its oppositions (see p.15) occurring a year and two weeks apart. When Saturn is viewed at a magnification of 100 times at opposition, its globe looks the same size as the full Moon seen with the naked eye. Saturn's disc is crossed by cloud belts, but these are not as prominent as those of Jupiter, nor is there an equivalent of the Great Red Spot. However, white spots appear in the equatorial zone about every 30 years, during summer in the planet's northern hemisphere.

Naked-eye view
Saturn is the bright, star-like object at the centre of this picture.

CCD view
With a CCD camera, it is possible to see the major divisions of the rings.

Telescope view
A small telescope shows the outline of Saturn's rings.

VIEWING SATURN'S RINGS
As Saturn orbits the Sun, we see its rings from various angles because of the combined effect of Saturn's axial tilt and the inclination of its orbit to our own. At most, the rings are tipped towards us at about 27 degrees. Twice in each orbit of Saturn around the Sun the rings appear to us edge-on, when they disappear from view because they are so thin. The next time this will happen will be in 2038 and 2039.

The Earth and Saturn in space

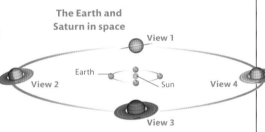

Appearance of rings as viewed from the Earth

View 1   View 2   View 3   View 4

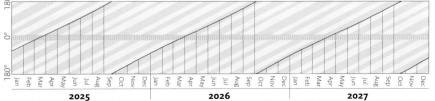

| Gravity (Earth = 1): 0.92 | Number of moons: 80+ | Average temperature: -176°C (-285°F) | Maximum magnitude: -0.3 |

STRUCTURE AND ATMOSPHERE

Saturn's cloudy atmosphere is similar to that of Jupiter, only colder and with an overlying haze, which gives Saturn its butterscotch colour and smoother look. Beneath the clouds is an interior of liquid hydrogen and helium, and a rocky core. Saturn's average density is only about 70 per cent that of water, the lowest of any planet. This low density, combined with its rapid rotation, gives it a highly elliptical shape, its equatorial diameter being nearly 12,000km (7,500 miles) greater than its polar diameter.

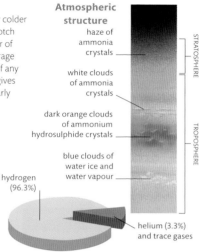

Atmospheric structure

- haze of ammonia crystals
- white clouds of ammonia crystals
- dark orange clouds of ammonium hydrosulphide crystals
- blue clouds of water ice and water vapour

STRATOSPHERE

TROPOSPHERE

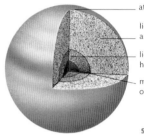

- atmosphere
- liquid hydrogen and helium
- liquid metallic hydrogen and helium
- massive core of rock and ice

Internal structure

hydrogen (96.3%)

helium (3.3%) and trace gases

Atmospheric composition

RINGS AND MOONS

Saturn's rings consist of an orbiting swarm of icy chunks, each no more than a few metres across. The orbits of the innermost moons lie among the rings. Pan orbits in a gap called the Encke Division in the outer part of Saturn's A ring. The next moon, Atlas, orbits at the edge of ring A, while Prometheus and Pandora orbit either side of ring F. Some moons actually share the same orbits.

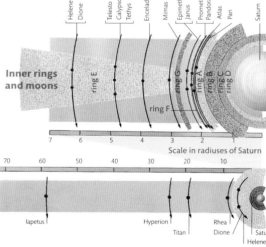

Inner rings and moons

Helene | Dione | Telesto | Calypso | Tethys | Enceladus | Mimas | Epimetheus | Janus | Prometheus | Pandora | Atlas | Pan | Saturn

ring E · ring F · ring G · ring A · ring B · ring C · ring D

7 6 5 4 3 2 1

Scale in radiuses of Saturn

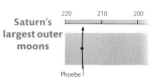

Saturn's largest outer moons

220 210 200 70 60 50 40 30 20 10

Phoebe | Iapetus | Hyperion | Titan | Rhea | Dione | Saturn | Helene

Location of Saturn 2028–2033

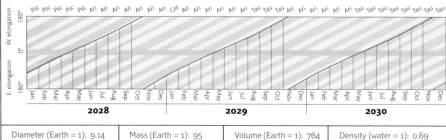

W. elongation 180°   E. elongation 180°   0°

Psc Psc Psc Psc Psc Psc Ari Ari Ari Ari Ari Ari Ari Cet Ari Ari Ari Ari Ari Tau Tau Tau Ari Ari Ari Ari Ari Ari Ari Tau Tau Tau Tau Tau Tau Tau Tau Tau

Jan Feb Mar Apr May Jun Jul Aug Sep Oct Nov Dec | Jan Feb Mar Apr May Jun Jul Aug Sep Oct Nov Dec | Jan Feb Mar Apr May Jun Jul Aug Sep Oct Nov Dec

2028   **2029**   **2030**

| Diameter (Earth = 1): 9.14 | Mass (Earth = 1): 95 | Volume (Earth = 1): 764 | Density (water = 1): 0.69 |

INNER RINGS

The outermost part of Saturn's ring system visible from the Earth is ring A, which is nearly 275,000km (170,000 miles) across. This is separated from the brightest and widest part, ring B, by the Cassini Division, a gap of about 4,500km (2,800 miles), visible with a 75mm (3in) telescope. Next is the partly transparent ring C (or crepe ring). Fainter rings, called D and F, lie inside and outside the visible rings. Two other very faint rings, known as G and E, lie beyond ring F but are not shown here.

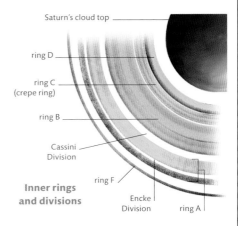

Saturn's cloud top

ring D

ring C
(crepe ring)

ring B

Cassini
Division

ring F

Inner rings
and divisions

Encke
Division

ring A

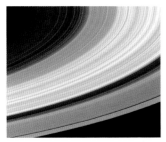

Cassini's view of Saturn's rings

Saturn's rings break up into thousands of tiny ringlets and gaps. This close-up image taken by NASA's Cassini probe shows the three main rings visible from Earth: the transparent inner C ring (top left), the bright central B ring, and the outer A ring.

MOONS

Saturn has over 80 known moons. The largest is Titan, 5,150km (3,200 miles) in diameter. It is the second-largest moon in the Solar System (after Jupiter's Ganymede) and the only one with a substantial atmosphere. At magnitude 8, it can be seen with a small telescope as it orbits the planet every 16 days. An aperture of 150mm (6in) should show several other moons – notably Rhea, Tethys, Dione, Iapetus, Enceladus, and perhaps Mimas. Iapetus is unusual in being over four times fainter when on one side of Saturn than on the other, because it has one bright and one dark hemisphere.

Titan

Titan's surface is obscured by orange hazes suspended in its nitrogen-rich atmosphere.

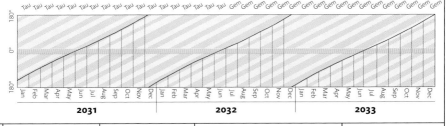

| Gravity (Earth = 1): 0.92 | Number of moons: 80+ | Average temperature: -176°C (-285°F) | Maximum magnitude: -0.3 |

URANUS

A COLD GAS GIANT, Uranus is the third-largest planet in the Solar System. Its most unusual characteristic is that its axis of rotation lies almost in the plane of its orbit, and so it seems to orbit the Sun on its side. Although it can be seen with the naked eye when at its brightest, Uranus was not discovered until 1781, when it was found by a British astronomer, William Herschel.

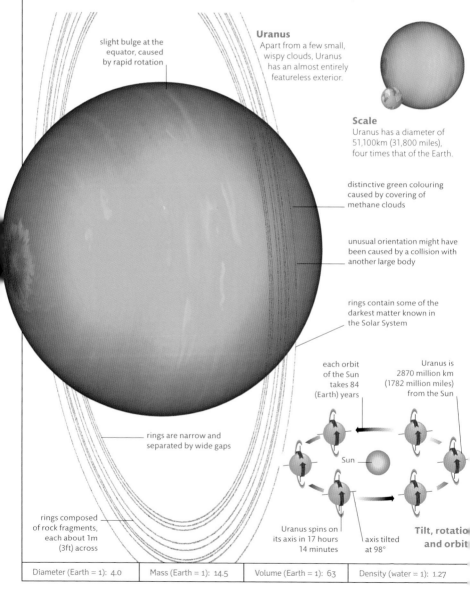

slight bulge at the equator, caused by rapid rotation

Uranus
Apart from a few small, wispy clouds, Uranus has an almost entirely featureless exterior.

Scale
Uranus has a diameter of 51,100km (31,800 miles), four times that of the Earth.

distinctive green colouring caused by covering of methane clouds

unusual orientation might have been caused by a collision with another large body

rings contain some of the darkest matter known in the Solar System

each orbit of the Sun takes 84 (Earth) years

Uranus is 2870 million km (1782 million miles) from the Sun

rings are narrow and separated by wide gaps

Sun

rings composed of rock fragments, each about 1m (3ft) across

Uranus spins on its axis in 17 hours 14 minutes

axis tilted at 98°

Tilt, rotation and orbit

| Diameter (Earth = 1): 4.0 | Mass (Earth = 1): 14.5 | Volume (Earth = 1): 63 | Density (water = 1): 1.27 |

ATMOSPHERE AND CLIMATE

Uranus is covered in methane clouds that absorb red light, giving the planet a greenish appearance. Its disc is nearly featureless, but some bright clouds were seen by the Voyager 2 space probe in 1986 and later by the Hubble Space Telescope. The planet's extreme axial tilt means that in the course of each orbit the Sun appears overhead at the equator and both poles; each pole experiences 42 years of daylight followed by 42 years of darkness. The interior is probably significantly different from those of Jupiter and Saturn, consisting mostly of water, methane, ammonia, and rock, rather than liquid hydrogen.

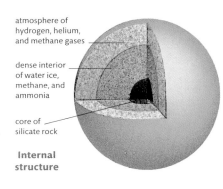

atmosphere of hydrogen, helium, and methane gases

dense interior of water ice, methane, and ammonia

core of silicate rock

Internal structure

Atmospheric composition

methane (2.3%) and trace gases

helium (15.2%)

hydrogen (82.5%)

Inner rings and moons

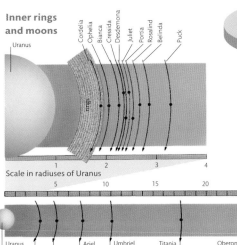

Uranus

Cordelia
Ophelia
Bianca
Cressida
Desdemona
Juliet
Portia
Rosalind
Belinda
Puck

rings

Scale in radiuses of Uranus

1   2   3   4

5   10   15   20

Uranus
Puck   Miranda   Ariel   Umbriel   Titania   Oberon

Outer moons

RINGS AND MOONS

Uranus has 27 known moons and 13 rings. The rings are too faint to be seen with the type of equipment used by amateur observers. Two of the rings are much farther from the planet than the others, almost constituting a ring system of their own. The brightest ring, called the Epsilon ring, is 100km (60 miles) wide and has "shepherd" moons, Cordelia and Ophelia, one on each side. The moons, too, are faint; even the biggest and brightest, Titania, is of only 14th magnitude and hence cannot be seen without a large telescope.

OBSERVATION

Under good conditions, and when its position is known, Uranus can be located with the naked eye, resembling a star of 6th magnitude. It is easy to find with binoculars, even from urban areas. Its identity can be confirmed by watching its movements from night to night. At oppositions, which occur a year and four days apart, a magnification of 500 times will show Uranus about the same size as the full Moon.

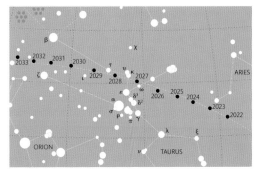

Location of Uranus 2022–2033

| Gravity (Earth = 1): 0.89 | Number of moons: 27 | Average temperature: -215°C (-355°F) | Maximum magnitude: 5.5 |

NEPTUNE

NEPTUNE IS THE outermost of the gas giants. Like Uranus, it has an atmosphere rich in hydrogen, helium, and methane, and a faint ring system. Neptune was discovered in 1846, after its position had been predicted based on disturbances to the motion of Uranus caused by its gravitational pull. The planet remained poorly known until the space probe Voyager 2 flew past in 1989.

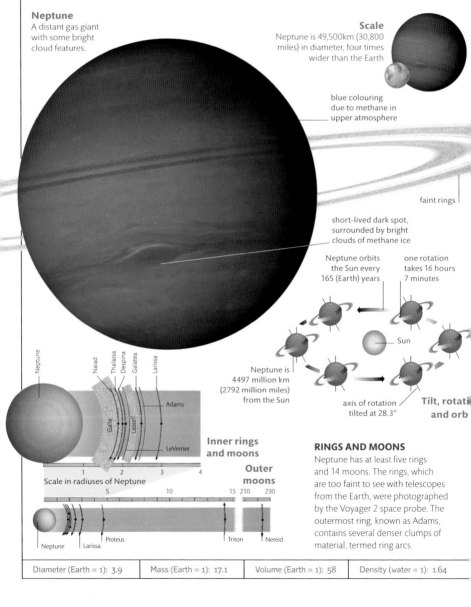

Neptune
A distant gas giant with some bright cloud features.

Scale
Neptune is 49,500km (30,800 miles) in diameter, four times wider than the Earth

blue colouring due to methane in upper atmosphere

faint rings

short-lived dark spot, surrounded by bright clouds of methane ice

Neptune orbits the Sun every 165 (Earth) years

one rotation takes 16 hours 7 minutes

Sun

Neptune is 4497 million km (2792 million miles) from the Sun

axis of rotation tilted at 28.3°

Tilt, rotation and orbit

Neptune

Naiad
Thalassa
Despina
Galatea
Larissa

Galle
Lassell
LeVerrier
Adams

Scale in radiuses of Neptune
1   2   3   4
5   10   15 210   230

Inner rings and moons

Outer moons

Neptune   Larissa   Proteus   Triton   Nereid

RINGS AND MOONS

Neptune has at least five rings and 14 moons. The rings, which are too faint to see with telescopes from the Earth, were photographed by the Voyager 2 space probe. The outermost ring, known as Adams, contains several denser clumps of material, termed ring arcs.

| Diameter (Earth = 1): 3.9 | Mass (Earth = 1): 17.1 | Volume (Earth = 1): 58 | Density (water = 1): 1.64 |

ATMOSPHERE AND CLIMATE

The atmospheres of Neptune and Uranus are similar, although Neptune's is stormier and appears blue (rather than blue-green) because there is more methane in its upper levels. In 1989 Voyager 2 photographed a large dark spot, reminiscent of Jupiter's Great Red Spot, in the southern hemisphere. Five years later, when the Hubble Space Telescope observed Neptune, the spot had vanished but another dark spot, this time in the northern hemisphere, appeared late in 1994. Bright streaks of methane cirrus clouds were recorded by both Voyager and Hubble. The composition of the main cloud deck is not known with certainty, but hydrogen sulphide and ammonia are probably present. Neptune's interior is thought to resemble that of Uranus, consisting of water, ammonia, and methane above a rocky core.

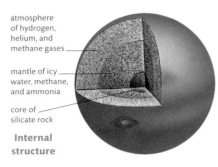

atmosphere of hydrogen, helium, and methane gases

mantle of icy water, methane, and ammonia

core of silicate rock

Internal structure

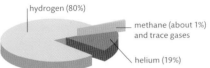

hydrogen (80%)

methane (about 1%) and trace gases

helium (19%)

Atmospheric composition

MOONS

Of Neptune's 14 known moons, Triton is the only one of any significance. It is also the largest, having a diameter of 2,700km (1,680 miles). The surface is covered with nitrogen and methane frost and, at -235°C (-391°F), is the coldest known place in the Solar System. Liquid nitrogen bursts through the surface to create geysers that leave dark streaks. Triton has a retrograde orbit (it moves in the opposite direction to Neptune), and is thought to have been a separate body, like Pluto, that was captured by Neptune's gravity. Triton is actually larger than Pluto, and the two may be physically similar.

Triton's south pole
The nitrogen and methane ice at Triton's poles has a slight but distinctive pink colouring.

OBSERVATION

Neptune never becomes brighter than 8th magnitude and so is invisible to the naked eye, but it can be located with binoculars, through which it resembles a faint star. Sightings of Neptune can be confirmed by observing its gradual movement against the background of stars over several nights. Through a telescope, the planet appears as a featureless bluish disc about two-thirds the apparent size of Uranus. Neptune comes to opposition (see p.15) only two-and-a-half days later each year.

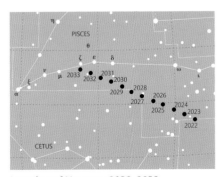

Location of Neptune 2022–2033

| Gravity (Earth = 1): 1.12 | Number of moons: 14 | Average temperature: -215°C (-355°F) | Maximum magnitude: 7.6 |

PLUTO AND THE OTHER DWARF PLANETS

IN THE DARK outer reaches of the Solar System beyond Neptune orbits a swarm of small icy bodies rather like a belt of outer asteroids. Collectively these are known as trans-Neptunian objects, or TNOs. Like comet nuclei, which they closely resemble, the TNOs are bits left over from the formation of the outer planets. The largest of these TNOs are classified as dwarf planets.

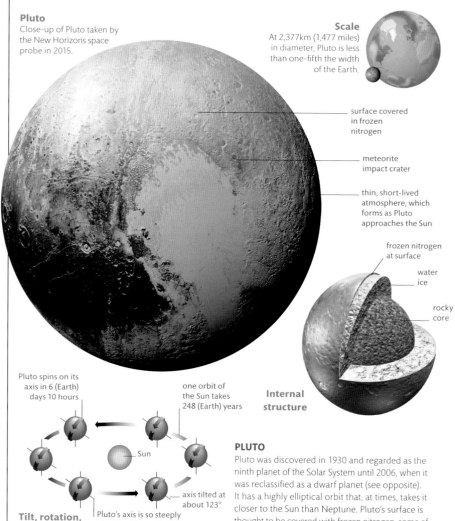

Pluto
Close-up of Pluto taken by the New Horizons space probe in 2015.

Scale
At 2,377km (1,477 miles) in diameter, Pluto is less than one-fifth the width of the Earth.

surface covered in frozen nitrogen

meteorite impact crater

thin, short-lived atmosphere, which forms as Pluto approaches the Sun

frozen nitrogen at surface

water ice

rocky core

Internal structure

Pluto spins on its axis in 6 (Earth) days 10 hours

one orbit of the Sun takes 248 (Earth) years

Sun

axis tilted at about 123°

Tilt, rotation, and orbit

Pluto's axis is so steeply tilted that its north pole lies below the plane of its orbit

PLUTO

Pluto was discovered in 1930 and regarded as the ninth planet of the Solar System until 2006, when it was reclassified as a dwarf planet (see opposite). It has a highly elliptical orbit that, at times, takes it closer to the Sun than Neptune. Pluto's surface is thought to be covered with frozen nitrogen, some of which evaporates to form a temporary atmosphere for a few decades when Pluto is closest to the Sun.

| Diameter (Earth = 1): 0.95 | Mass (Earth = 1): 0.002 | Volume (Earth = 1): 0.006 | Density (water = 1): 2.0 |

CHARON

Pluto has five moons, the largest of which, Charon, was discovered in 1978. It has about half the diameter and one-eighth the mass of Pluto, making it the largest moon in relation to its parent in the Solar System. In effect, the two bodies form a double planet. Charon orbits about 18,400km (11,400 miles) above Pluto's surface every 6.4 days. Pluto and Charon each spin on their axes in 6.4 days as well, so each body always presents the same face to the other.

Charon

a spot drawn on Pluto's surface faces a spot on Charon

as Pluto and Charon rotate, the two spots still face each other

Synchronized orbits

From one hemisphere of Pluto, Charon is always visible and remains motionless in the sky. It can never be seen from the other hemisphere.

Charon is never visible from this side of Pluto

Physical properties of Pluto and Eris

Property	Pluto	Eris
Diameter (km)	2,377	2,326
Perihelion (million km)	4,425	5,700
Aphelion (million km)	7,375	14,600
Orbital period (years)	247.9	560
Orbital eccentricity	0.25	0.44
Inclination	17.1°	44°
Number of moons	5	1

DWARF PLANETS

In 2005 astronomers discovered an object in the Solar System which is almost as large as Pluto. It is now named Eris. This led to the introduction in 2006 of a new category of Solar System objects, called dwarf planets. These are bodies that have enough mass and gravity to make them roughly spherical but are not big enough to have cleared other objects out of their path, which the major planets have done. Pluto is now classified as a dwarf planet, as is Eris. This category also includes Ceres, the largest member of the asteroid belt (see p.62). Two other trans-Neptunian objects, Haumea and Makemake, were added to the dwarf planet category in 2008.

Eris and Dysnomia

In September 2005, astronomers using the 10-metre Keck telescope in Hawaii discovered a moon orbiting Eris (seen to the left of Eris in this image from the Hubble Space Telescope). It is named Dysnomia, after the daughter of Eris – the Greek goddess of discord.

| Gravity (Earth = 1): 0.07 | Number of moons: 1 | Average temperature: -223°C (-369°F) | Maximum magnitude: 14 |

COMETS AND METEORS

COMETS, consisting of frozen gas and dust, exist in a swarm at the edge of the Solar System. Sometimes they approach the Sun on highly elongated orbits, warming up and releasing gas and dust, before receding back into the darkness. Dust particles from comets can enter the Earth's atmosphere, where they burn up to create bright streaks in the sky called meteors or shooting stars.

COMETS

Comets are icy remnants from the formation of the outer planets. They usually orbit unseen in the so-called Oort Cloud, which extends to one light year or more from the Sun. Sometimes the gravity of a passing star nudges a comet from the cloud into the inner Solar System, so that it becomes visible from Earth. Thousands of comets are known, but the Oort Cloud and its inner region, called the Kuiper Belt, are thought to contain billions of them.

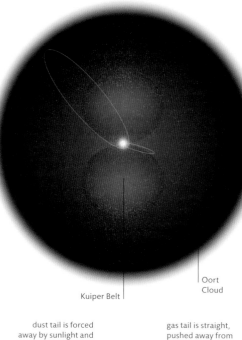

The sources of comets
Comets of long orbital period are thought to come from the Oort Cloud (at the edge of this picture); those of shorter period come from the Kuiper Belt (the lighter blue area).

Oort Cloud

Kuiper Belt

NUCLEI AND TAILS

The only solid part of a comet is its nucleus, which is typically about a kilometre across. As it approaches the Sun, the nucleus warms up, emitting gas and dust to form a glowing head, termed the coma, up to 100,000km (60,000 miles) across. In some comets, gas and dust from the coma form tails.

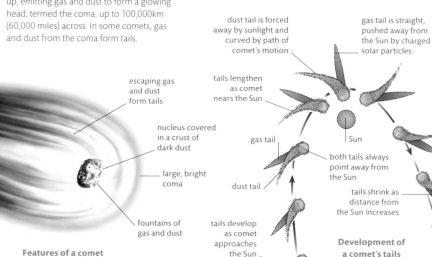

escaping gas and dust form tails

nucleus covered in a crust of dark dust

large, bright coma

fountains of gas and dust

Features of a comet

dust tail is forced away by sunlight and curved by path of comet's motion

gas tail is straight, pushed away from the Sun by charged solar particles

tails lengthen as comet nears the Sun

gas tail

dust tail

Sun

both tails always point away from the Sun

tails shrink as distance from the Sun increases

tails develop as comet approaches the Sun

Development of a comet's tails

COMET OBSERVATION

Dozens of comets can be seen each year with a large telescope, but only rarely is one bright enough to be visible to the naked eye. There are two main classes: periodic comets orbit the Sun in under 200 years, while long-period comets may take hundreds of thousands of years to return. Large, bright comets develop tails, of which there are two types: one consists of gas and looks bluish; the other consists of dust and appears yellowish. Gas tails are usually narrow and straight, while dust tails are often broad and curved. Both are visible through binoculars.

Comet Hale-Bopp
This bright comet was easily visible to the naked eye in 1997. It will not return for 2,400 years.

The origin of a meteor shower
Individual meteor showers occur at the same time each year as the Earth's orbit passes through the trail of dust left along a comet's orbit.

METEORS

On any clear night bright streaks can be seen in the sky from time to time. They appear suddenly and last for less than a second. These are meteors, or shooting stars, caused by specks of dust from comets burning up in the atmosphere at a height of about 100km (60 miles). A handful of so-called sporadic (random) meteors can usually be seen in an hour, but several times each year the Earth passes through a trail of dust left by a comet, resulting in a meteor shower. Meteors in a shower appear to diverge from a single point called the radiant, and the shower is named after the constellation in which the radiant is found – hence, the Leonids appear to come from Leo. Meteors are easy to see with the naked eye, and useful data have been derived from the records kept by amateur observers.

SHOWER	DATE	CONSTELLATION	HOURLY RATE
Quadrantids	3–4 January	Boötes	100
Lyrids	21 April	Lyra	10
Eta Aquariids	5 May	Aquarius	35
Delta Aquariids	29 July	Aquarius	25
Perseids	12 August	Perseus	80
Orionids	20–22 October	Orion	25
Taurids	5 November	Taurus	10
Leonids	17 November	Leo	10
Geminids	13 December	Gemini	100
Ursids	23 December	Ursa Minor	10

Bright meteor showers
This table shows the main meteor showers that the Earth encounters each year. The hourly rate is the maximum number visible under ideal conditions. The date given indicates the time of maximum activity – smaller numbers of meteors can usually be seen for some days either side of this date.

ASTEROIDS AND METEORITES

ASTEROIDS – or minor planets – are small bodies of iron and rock left over from the formation of the Solar System. Most asteroids orbit the Sun in a belt between Mars and Jupiter, but some stray from this group to cross the paths of the inner planets. Meteorites are small fragments, usually from asteroids, that have entered the Earth's atmosphere and fallen to the surface.

ASTEROIDS

Well over a million asteroids are known, and thousands more are being discovered each year. Most asteroids orbit in the belt between Mars and Jupiter, but there is a group, known as the Trojan asteroids, that moves along Jupiter's orbit. Members of another group, named the Apollo asteroids, cross the Earth's orbit and may therefore collide with our planet. The largest asteroid, known as Ceres, is 940km (580 miles) in diameter. The brightest is Vesta, 525km (325 miles) across, which can be as bright as 5th magnitude, making it easily visible with binoculars.

Ida
This asteroid, 55km (34 miles) in length, is orbited by a small satellite known as Dactyl.

Trojan asteroids

asteroid belt

orbit of an Apollo asteroid

Sun

Earth

Venus

Mars

Mercury

Jupiter

Trojan asteroids

Asteroid orbits
The asteroid belt lies between about 320 and 490 million km (200 and 305 million miles) from the Sun. The Trojans orbit either side of Jupiter.

METEORITES

Every day about 10 meteorites fall to Earth, although most land in remote areas or the oceans and go unnoticed. Over 60,000 meteorites have been found, many long after their fall. Meteorites are composed of rock, or iron, or a mixture of both. Large meteorites are usually moving fast enough to dig a crater when they hit the Earth, but smaller ones are slowed by the atmosphere and drop harmlessly to the surface.

Meteor crater
This crater, 1.2km (0.75 miles) wide, was formed in the desert of Arizona, USA, by the impact of an iron meteorite about 50,000 years ago.

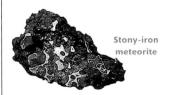

Stony-iron meteorite

THE CONSTELLATIONS

HOW THIS SECTION WORKS

THIS SECTION contains entries on the 88 constellations, arranged alphabetically by their internationally accepted names. The charts are the most detailed in the book, showing all stars above the approximate threshold of naked-eye visibility. The text covers the origin of the constellations and a selection of the stars and deep-sky objects that they contain.

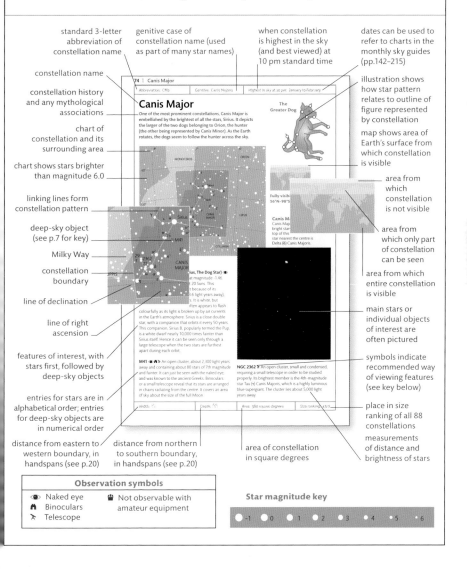

standard 3-letter abbreviation of constellation name

genitive case of constellation name (used as part of many star names)

when constellation is highest in the sky (and best viewed) at 10 pm standard time

dates can be used to refer to charts in the monthly sky guides (pp.142–215)

constellation name

constellation history and any mythological associations

chart of constellation and its surrounding area

chart shows stars brighter than magnitude 6.0

linking lines form constellation pattern

deep-sky object (see p.7 for key)

Milky Way

constellation boundary

line of declination

line of right ascension

features of interest, with stars first, followed by deep-sky objects

entries for stars are in alphabetical order; entries for deep-sky objects are in numerical order

distance from eastern to western boundary, in handspans (see p.20)

distance from northern to southern boundary, in handspans (see p.20)

illustration shows how star pattern relates to outline of figure represented by constellation

map shows area of Earth's surface from which constellation is visible

area from which constellation is not visible

area from which only part of constellation can be seen

area from which entire constellation is visible

main stars or individual objects of interest are often pictured

symbols indicate recommended way of viewing features (see key below)

place in size ranking of all 88 constellations

measurements of distance and brightness of stars

area of constellation in square degrees

Contents of chart entry (Canis Major example)

74 | Canis Major

| Abbreviation: CMa | Genitive: Canis Majoris | Highest in sky at 10 pm: January to February |

Canis Major

The Greater Dog

One of the most prominent constellations, Canis Major is embellished by the brightest of all the stars, Sirius. It depicts the larger of the two dogs belonging to Orion, the hunter (the other being represented by Canis Minor). As the Earth rotates, the dogs seem to follow the hunter across the sky.

Fully visible 56°N–90°S

Canis Major... bright star... top of this... star nearest the centre is Delta (δ) Canis Majoris.

Sirius, The Dog Star At magnitude –1.46... 20 Suns. This... because of its 8.6 light years away), s. It is white, but often appears to flash colourfully as its light is broken up by air currents in the Earth's atmosphere. Sirius is a close double star, with a companion that orbits it every 50 years. This companion, Sirius B, popularly termed the Pup, is a white dwarf nearly 10,000 times fainter than Sirius itself. Hence it can be seen only through a large telescope when the two stars are furthest apart during each orbit.

M41 An open cluster, about 2,300 light years away and containing about 80 stars of 7th magnitude and fainter. It can just be seen with the naked eye, and was known to the ancient Greeks. Binoculars or a small telescope reveal that its stars are arranged in chains radiating from the centre. It covers an area of sky about the size of the full Moon

NGC 2362 An open cluster, small and condensed, requiring a small telescope in order to be studied properly. Its brightest member is the 4th-magnitude star Tau (τ) Canis Majoris, which is a highly luminous blue supergiant. The cluster lies about 5,000 light years away.

| Width: | Depth: | Area: 380 square degrees | Size ranking: 43/88 |

Abbreviation: And	Genitive: Andromedae	Highest in sky at 10 pm: October to November

Andromeda

One of the most famous constellations, Andromeda depicts the princess in Greek myth who was chained to a rock as a sacrifice to a sea monster but was saved by the hero Perseus. The constellation contains the nearest major galaxy to us, M31, the Andromeda Galaxy, which is the most distant object visible to the naked eye.

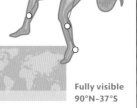

Andromeda

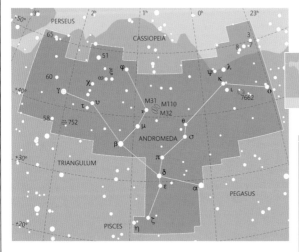

**Fully visible
90°N–37°S**

**M31 (The
Andromeda Galaxy)**
This spiral galaxy looks elliptical because it is tilted at an angle to us.

Around M31
M31 is at the top of this picture. The bright star at the bottom left is Beta (β) Andromedae.

Moon diameters. Even so, this is only the central part of the galaxy, and a larger telescope is needed to see any trace of its spiral arms. Lying about 2.5 million light years away, M31 is the largest galaxy in the Local Group. It has two small elliptical companion galaxies (the equivalent of our Magellanic Clouds), M32 and NGC 205 (or M110), which can be seen with a telescope with an aperture of 100mm (4in) or more.

FEATURES OF INTEREST

Gamma (γ) Andromedae ✶ One of the most attractive double stars in the sky. It appears to the naked eye as a single star of magnitude 2.1, but even a small telescope will reveal an orange-coloured primary, which is a giant star, and a blue companion of 5th magnitude.

M31 (The Andromeda Galaxy) ◉ ♁ ✶
A spiral galaxy, similar to the Milky Way but larger. On a clear, dark night, it can be seen by the naked eye as an elongated smudge. With binoculars or a small telescope, it looks much larger, extending for several

NGC 752 ♁ An open cluster, visible through binoculars, consisting of faint stars spread over an area of sky wider than the full Moon. It lies about 1,300 light years away.

NGC 7662 ✶ A planetary nebula, about 4,000 light years away. It looks small but prominent through a telescope, like a blue-green star of 9th magnitude; high magnification reveals a rounded outline, like an out-of-focus star.

Width:	Depth:	Area: 722 square degrees	Size ranking: 19th

Abbreviation: Ant	Genitive: Antliae	Highest in sky at 10 pm: March to April

Antlia

Antlia is a faint constellation, considerably overshadowed by its glorious southern neighbours Centaurus and Vela. Antlia was invented in the 18th century by the French astronomer Nicolas Louis de Lacaille and represents a mechanical air pump. Its brightest star is Alpha (α) Antliae, magnitude 4.3.

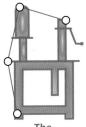

The Air Pump

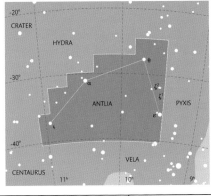

**Fully visible
49°N–90°S**

FEATURES OF INTEREST

Zeta (ζ) Antliae ♠ ☆ A multiple star that appears as a wide 6th-magnitude double when viewed through binoculars. When a small telescope is turned on it, the brighter star of the pair is seen also to be double. All three stars lie about 350 light years away.

Width:	Depth:	Area: 239 square degrees	Size ranking: 62nd

Abbreviation: Aps	Genitive: Apodis	Highest in sky at 10 pm: May to July

Apus

Apus lies near the south celestial pole, to the south of Centaurus. It is not easy to identify and contains few objects of note. It is one of the constellations invented at the end of the 16th century by the Dutch navigators Pieter Dirkszoon Keyser and Frederick de Houtman, and represents a bird of paradise of New Guinea. Its brightest star is Alpha (α) Apodis, magnitude 3.8.

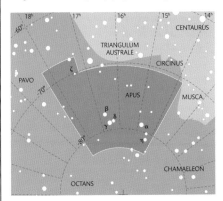

The Bird of Paradise

**Fully visible
7°N–90°S**

FEATURES OF INTEREST

Delta (δ) Apodis ♠ ☆ A wide double star, easily divided with binoculars. The two stars are of magnitudes 4.7 and 5.3. They are both red giants but are estimated to lie at different distances from us, 760 and 550 light years respectively, so they probably form an optical double rather than a genuine binary.

Width:	Depth:	Area: 206 square degrees	Size ranking: 67th

Abbreviation: Aqr	Genitive: Aquarii	Highest in sky at 10 pm: August to October

Aquarius

This well-known constellation represents a youth pouring water from a jar. The "Water-jar" is represented by a Y-shaped group of four stars, Gamma (γ), Zeta (ζ), Eta (η), and Pi (π) Aquarii. The stream of water flows into the mouth of a large fish, represented by the constellation of Piscis Austrinus to the south. Aquarius is a constellation of the zodiac, the Sun passing through it from 16 February to 11 March.

The Water Carrier

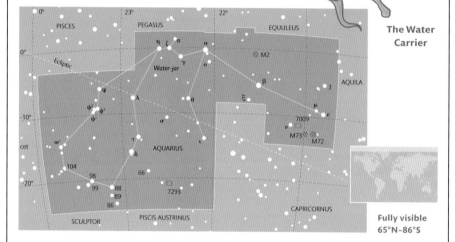

Fully visible 65°N–86°S

FEATURES OF INTEREST

Zeta (ζ) Aquarii ✶ A close binary, consisting of two white, 4th-magnitude stars that orbit each other every 430 years or so. Using high magnification, a telescope with an aperture of 75mm (3in) or more should separate them.

M2 ♠ ✶ A globular cluster, just too faint to be seen with the naked eye in all but the best conditions, but easy to see with binoculars or a small telescope. It resembles a fuzzy star.

NGC 7009 (The Saturn Nebula) ✶
A planetary nebula. When seen through a telescope with an aperture of 200mm (8in) or more, it appears to have appendages that resemble the rings of Saturn, hence its popular name. A smaller telescope will show it as an 8th-magnitude disc of similar apparent size to the globe of Saturn.

NGC 7293 (The Helix Nebula) ♠ ✶
Possibly the closest planetary nebula to Earth (only about 650 light years away), and one of the largest in apparent size, at over a third the diameter of the full Moon. Being so large, its light is spread over a wide area, making it difficult to spot. Under clear, dark skies, it can be seen through binoculars or a wide-field telescope as a pale grey patch.

NGC 7293 (The Helix Nebula)
On colour photographs, the Helix Nebula resembles an exotic flower. The shell of gas has been discarded from the faint star at the centre of the nebula.

Width:	Depth:	Area: 980 square degrees	Size ranking: 10th

Abbreviation: Aql	Genitive: Aquilae	Highest in sky at 10 pm: July to August

Aquila

The Eagle

Aquila lies on the celestial equator and represents the eagle that carried the thunderbolts of the Greek god Zeus. The Milky Way passes through the constellation, and there are dense star fields towards the border with Scutum; the brightest part is an area known as the Scutum Star Cloud.

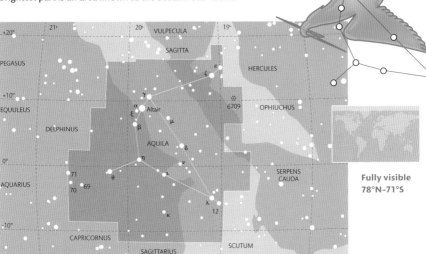

**Fully visible
78°N–71°S**

FEATURES OF INTEREST

Alpha (α) Aquilae (Altair)  Among the 20 brightest stars in the sky, at magnitude 0.8, and also one of the closest first-magnitude stars to us, only 17 light years away. It marks the neck of the eagle and forms one corner of the Summer Triangle, completed by Deneb (in Cygnus) and Vega (in Lyra). Altair is flanked by two stars, the 4th-magnitude Beta (β) Aquilae (also known as Alshain) and 3rd-magnitude Gamma (γ) Aquilae (or Tarazed).

Eta (η) Aquilae A Cepheid variable star, one of the brightest in this important class. Its magnitude varies from 3.5 to 4.3 in a cycle lasting 7 days 4 hours, and the changes can be followed with the naked eye or through binoculars. It is a supergiant star, estimated to lie about 1,400 light years away.

The border between Aquila and Scutum
This distinctive hook-shaped group of stars is in the southern part of Aquila. The brightest star in the group (centre left) is Lambda (λ) Aquilae, magnitude 3.4.

The main stars in Aquila
Altair is the bright star at the top left of this picture, flanked on its left and right by Beta (β) and Gamma (γ) Aquilae, also known as Alshain and Tarazed.

Width:	Depth:	Area: 652 square degrees	Size ranking: 22nd

Abbreviation: Ara	Genitive: Arae	Highest in sky at 10 pm: June to July

Ara

Ara lies in the Milky Way, south of Scorpius. It is well to the south of the celestial equator but was known to the ancient Greeks, who visualized it as the altar on which their gods swore an oath of allegiance before challenging the Titans for control of the Universe. It is also depicted as the altar on which Centaurus is about to sacrifice Lupus, the wolf.

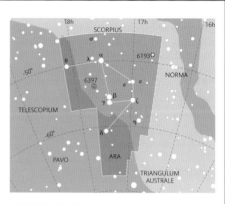

The Altar

Fully visible 22°N–90°S

FEATURES OF INTEREST

NGC 6193 🌑 An open cluster, consisting of a handful of stars scattered over an area half the apparent size of the full Moon. Its brightest star is of 6th magnitude, and the whole cluster is best seen with binoculars.

NGC 6397 🌑✶ A large globular cluster, about two-thirds the apparent size of the full Moon and one of the closest to us, at about 7,800 light years away. It is just visible to the naked eye and easy to find with binoculars. The stars, less densely packed than in many globulars, form chains in the outer regions that can be seen with a moderate-sized telescope.

Width: ✋	Depth: 🤚🤚	Area: 237 square degrees	Size ranking: 63rd

---

Abbreviation: Ari	Genitive: Arietis	Highest in sky at 10 pm: November to December

Aries   The Ram

Aries depicts the ram with the golden fleece, famous from Greek mythology. Its only noticeable feature is a line of three stars: Alpha (α), Beta (β), and Gamma (γ) Arietis. It is a constellation of the zodiac, the Sun passing through it from 18 April to 14 May. Aries lies between Pisces and Taurus.

The central part of Aries
Alpha (α), Beta (β), and Gamma (γ) Arietis are to the right in this picture.

FEATURES OF INTEREST

Gamma (γ) Arietis ✶ A wide double star, easily divided with a small telescope. The two stars are nearly identical, both being blue-white and of magnitude 4.6. Found by the English scientist Robert Hooke in 1664, when telescopes were still quite crude, this was one of the first double stars to be discovered.

Fully visible 90°N–58°S

Width: ✋🤚	Depth: 🤚🤚	Area: 441 square degrees	Size ranking: 39th

Abbreviation: Aur	Genitive: Aurigae	Highest in sky at 10 pm: December to February

Auriga

The Charioteer

Auriga represents the driver of a horse-drawn chariot. According to one myth, he is Erichthonius, a legendary king of Athens. However, there is no explanation in mythology for his depiction in the sky with a goat and its kids on his left arm. The goat is marked by the constellation's brightest star, Capella (a Latin name, meaning "she-goat"), while the kids (also known as the Haedi, another Latin name) are depicted by Zeta (ζ) and Eta (η) Aurigae. In Greek and Roman times, the figure's right foot was represented by a star now assigned to Taurus, Beta (β) Tauri.

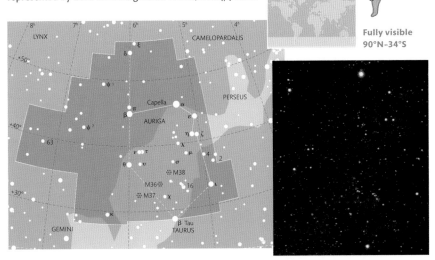

Fully visible 90°N–34°S

The main stars in Auriga
Capella is at the top of this picture. M36, M37, and M38 are just visible.

FEATURES OF INTEREST

Alpha (α) Aurigae (Capella) ◉ The sixth-brightest star in the sky, at magnitude 0.1. To the naked eye, it appears yellowish. It is, in fact, a spectroscopic binary, consisting of two yellow-coloured giants that orbit each other every 104 days. It lies 43 light years away.

Epsilon (ε) Aurigae ◉ One of the most extraordinary variable stars in the sky. It is an eclipsing binary, consisting of a brilliant white supergiant orbited by an odd dark companion that passes in front of it every 27 years, the longest known period between eclipses of any variable star. The star's brightness is more than halved by the eclipse, from magnitude 2.9 to 3.8, and it remains dimmed for over a year. From observations of the last eclipse, which lasted from 2009 into 2011, astronomers have concluded that the mystery partner is a hot, blue star obscured from view by a large disk of dark dust and gas seen almost edge-on.

Zeta (ζ) Aurigae ◉ An eclipsing binary, comprising an orange giant orbited every 2.7 years by a smaller blue star. During an eclipse, which lasts 40 days, the star's brightness drops from magnitude 3.7 to 4.0.

M36, M37, and M38 🔭 Just visible with the naked eye and easy to see with binoculars, these three open clusters lie 4,000–4,500 light years away. In a binocular field of 6 degrees or more, all three can be seen as misty patches. The smallest of the trio, M36, is the easiest to spot, a small telescope resolving its brightest stars. M37, the largest of the clusters at about two-thirds the width of the full Moon, contains more stars but they are fainter. M38 is the most scattered cluster; a small telescope reveals that many of its stars form chains.

Width:	Depth:	Area: 657 square degrees	Size ranking: 21st

Abbreviation: Boo	Genitive: Boötis	Highest in sky at 10 pm: May to June

Boötes

The Herdsman

Boötes is an elongated constellation depicting a man herding a bear, represented by Ursa Major. The name of its brightest star, Arcturus, is Greek for "bear guard". The northern part of Boötes contains the faint stars that formed the now-defunct constellation of Quadrans Muralis, the mural (or wall) quadrant, which gave its name to the Quadrantid meteor shower that radiates from this area every January.

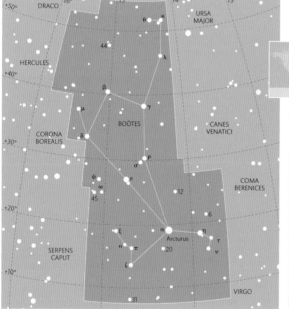

Fully visible
90°N–35°S

The main stars in Boötes
The bright star at the bottom of this picture is Arcturus.

FEATURES OF INTEREST

Alpha (α) Boötis (Arcturus) ◑ The brightest star north of the celestial equator, and, at magnitude -0.1, the fourth brightest overall. It is a red giant, 100 times more luminous than the Sun but with a cooler (and hence redder) surface. To the naked eye, it has a noticeably warm tint, which becomes stronger when it is viewed with binoculars.

Epsilon (ε) Boötis ⚹ A beautiful double star but a difficult one to divide. To the naked eye, it appears of magnitude 2.4. A telescope of aperture 75mm (3in) or more shows a red-giant primary and a close, blue-green partner of 5th magnitude. High magnification and a steady night are needed to separate the stars.

Kappa (κ) Boötis ⚹ A double star, easily divided with a small telescope into a pair of white stars of 5th and 7th magnitudes.

Mu (μ) Boötis ♦⚹ A multiple star. Binoculars show a wide double, with stars of magnitudes 4.3 and 6.5. The fainter star is itself double, with 7th- and 8th-magnitude components that orbit each other every 260 years. They should be divisible in moderate apertures, but are just too close to be split with a small telescope.

Xi (ξ) Boötis ⚹ A double star, divisible with a small telescope. The stars, of magnitudes 4.7 and 7.0, are both yellow-orange, one slightly deeper in tone than the other. They form a true binary with an orbital period of 152 years.

Width:	Depth:	Area: 907 square degrees	Size ranking: 13th

Abbreviation: Cae	Genitive: Caeli	Highest in sky at 10 pm: December to January

Caelum

The Chisel

Caelum is a small, faint constellation in an inconspicuous area of the southern-hemisphere sky. It represents a stonemason's chisel and was introduced into the sky in the 18th century by the French astronomer Nicolas Louis de Lacaille. Its brightest star, Alpha (α) Caeli, is of magnitude 4.4.

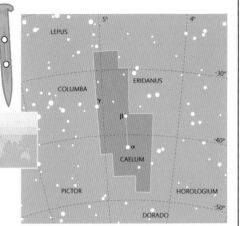

Fully visible
41°N–90°S

FEATURES OF INTEREST

Gamma (γ) Caeli ⚹ A double star, although a moderate-sized telescope is needed to show it as such. It is of magnitude 4.6, with a close companion of 8th magnitude.

Width: 🖐	Depth: 🖐🖐	Area: 125 square degrees	Size ranking: 81st

Abbreviation: Cam	Genitive: Camelopardalis	Highest in sky at 10 pm: December to May

Camelopardalis

This faint constellation of the northern sky, representing a giraffe, was introduced in the early 17th century by the Dutch astronomer Petrus Plancius. An obsolete variant of its name is Camelopardus.

The Giraffe

Fully visible
90°N–3°S

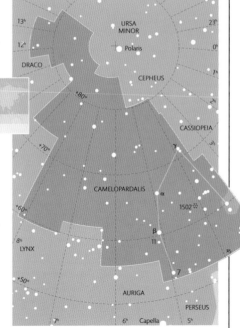

FEATURES OF INTEREST

Beta (β) Camelopardalis 🔭⚹
A wide double star. The brighter component is the brightest star in the constellation, at magnitude 4.0. It has a companion of magnitude 7.4 that is easy to see with a small telescope, or even a pair of powerful binoculars. In the same binocular field, an even wider pairing can be seen; the two stars, 11 and 12 Camelopardalis, are of magnitudes 5.2 and 6.1.

NGC 1502 🔭⚹ An open cluster, visible with binoculars or a small telescope. Although small, its stars are moderately bright, and include a noticeable 7th-magnitude double.

Width: 🖐🖐	Depth: 🖐🖐	Area: 757 square degrees	Size ranking: 18th

| Abbreviation: Cnc | Genitive: Cancri | Highest in sky at 10 pm: February to March |

Cancer

Cancer represents the crab that, in Greek mythology, was crushed underfoot by Hercules during his battle with the multi-headed Hydra. It lies between Gemini and Leo and is the faintest constellation of the zodiac; its brightest star, Beta (β) Cancri, is of magnitude 3.5. The Sun is within Cancer's boundaries from 20 July until 10 August.

The Crab

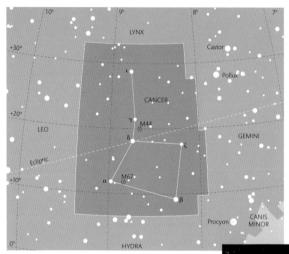

Fully visible
90°N–57°S

M44 (Praesepe)
M44 is too large to fit in the field of view of most telescopes. This picture shows the cluster as it looks through binoculars. The two stars known as the aselli that are useful in locating the cluster are just out of view.

FEATURES OF INTEREST

Zeta (ζ) Cancri ✶ A multiple star. Through a small telescope, it is seen to consist of two stars, of magnitudes 5.1 and 6.2. A telescope with an aperture larger than about 150mm (6in) will show that the brighter component has a much closer companion, of magnitude 6.1, which orbits it every 60 years.

Iota (ι) Cancri ✶ A wide double star, with components of magnitudes 4.0 and 6.6 that are easy to divide with a small telescope.

M44 (Praesepe) ♙ An open cluster, also known as the Beehive Cluster or the Manger. (Praesepe is Latin, meaning both "manger" and "hive".) It appears as a cloudy patch at the limit of naked-eye visibility – its brightest stars are of 6th magnitude, and it was known to the ancient Greeks – but binoculars show it as a field of stars more than three times the apparent width of the full Moon. It lies about 600 light years away. To the north and south of the cluster are the stars Gamma (γ) Cancri (magnitude 4.7) and Delta (δ) Cancri (magnitude 3.9). In ancient

times, these stars were visualized as donkeys feeding at the manger, hence they are known as the aselli, or asses.

M67 ♙✶ An open cluster. It contains more stars than M44 (Praesepe), but it is further away from us (about 2,600 light years) and so appears fainter and smaller.

| Width: 🖐🖐 | Depth: 🖐🖐 | Area: 506 square degrees | Size ranking: 31st |

Abbreviation: CVn	Genitive: Canum Venaticorum	Highest in sky at 10 pm: April to May

Canes Venatici

Canes Venatici represents two hunting dogs, held on a leash by the herdsman Boötes. The constellation was formed in 1687 by Johannes Hevelius from stars that had formerly been part of Ursa Major.

The
Hunting Dogs

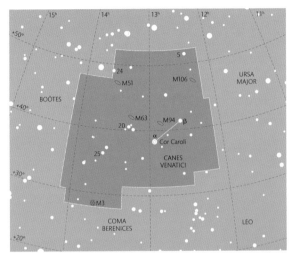

Fully visible
90°N–37°S

M51 (The Whirlpool Galaxy)
This spiral galaxy is interacting with a smaller companion galaxy, NGC 5195. M51 appears to us face-on, with NGC 5195 at the end of one of its spiral arms.

FEATURES OF INTEREST

Alpha (α) Canum Venaticorum (Cor Caroli) ✶
A double star. A small telescope will easily separate the two components, of magnitudes 2.9 and 5.6. Although both should appear white, some observers have reported delicate colour tints. This is possibly because the brighter star has an unusual composition, as revealed by spectroscopic studies. The name Cor Caroli means "Charles's heart", and commemorates King Charles I of England.

M3 ♠✶ A globular cluster on the verge of naked-eye visibility. It is easy to find with binoculars or a small telescope, in which it appears about half the size of the full Moon. A telescope with an aperture of about 100mm (4in) or more is needed to see the cluster's individual stars.

M51 (The Whirlpool Galaxy) ♠✶ A spiral galaxy, seen almost face-on. This is one of the most famous galaxies in the sky, and one of the easiest in which to see spiral structure. The galaxy can be located using binoculars. A small telescope reveals its bright nucleus, as well as the nucleus of a smaller, irregularly shaped galaxy, NGC 5195, which is passing close by. Using a

telescope of moderate to large aperture – 200mm (8in) or more – it should be possible to trace the outline of M51's arms. M51 and NGC 5195 are both about 25 million light years away.

Width:	Depth:	Area: 465 square degrees	Size ranking: 38th

Abbreviation: CMa	Genitive: Canis Majoris	Highest in sky at 10 pm: January to February

Canis Major

One of the most prominent constellations, Canis Major is embellished by the brightest of all the stars, Sirius. It depicts the larger of the two dogs belonging to Orion, the hunter (the other being represented by Canis Minor). As the Earth rotates, the dogs seem to follow the hunter across the sky.

The Greater Dog

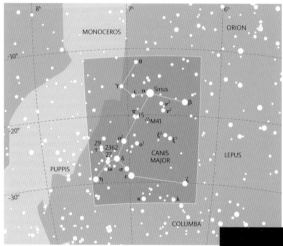

Fully visible
56°N–90°S

Canis Major
Canis Major contains several bright stars. Sirius is near the top of this picture; the bright star nearest the centre is Delta (δ) Canis Majoris.

FEATURES OF INTEREST

Alpha (α) Canis Majoris (Sirius, The Dog Star) ◉
The brightest star in the sky, at magnitude -1.46. Sirius emits the light of about 20 Suns. This luminosity is not unusual, but because of its relative closeness to us (it is 8.6 light years away), Sirius outshines all other stars. It is white, but when low on the horizon it often appears to flash colourfully as its light is broken up by air currents in the Earth's atmosphere. Sirius is a close double star, with a companion that orbits it every 50 years. This companion, Sirius B, popularly termed the Pup, is a white dwarf nearly 10,000 times fainter than Sirius itself. Hence it can be seen only through a large telescope when the two stars are furthest apart during each orbit.

M41 ◉♁✸ An open cluster, about 2,300 light years away and containing about 80 stars of 7th magnitude and fainter. It can just be seen with the naked eye, and was known to the ancient Greeks. Binoculars or a small telescope reveal that its stars are arranged in chains radiating from the centre. It covers an area of sky about the size of the full Moon.

NGC 2362 ✸ An open cluster, small and condensed, requiring a small telescope in order to be studied properly. Its brightest member is the 4th-magnitude star Tau (τ) Canis Majoris, which is a highly luminous blue supergiant. The cluster lies about 5,000 light years away.

Width:	Depth:	Area: 380 square degrees	Size ranking: 43rd

| Abbreviation: CMi | Genitive: Canis Minoris | Highest in sky at 10 pm: February |

Canis Minor

The
Lesser Dog

Canis Minor represents the smaller of the two dogs of Orion, the hunter. Both Orion and Canis Major, which depicts the larger dog, lie nearby. Apart from its two brightest stars, Canis Minor contains little of note.

**Fully visible
89°N–77°S**

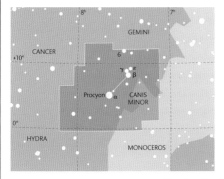

FEATURES OF INTEREST
Alpha (α) Canis Minoris (Procyon)  The 8th-brightest star in the sky, at magnitude 0.4. The name Procyon is of Greek origin, meaning "before the dog", from the fact that it rises before the other dog star, Sirius. It is a white star, slightly less luminous than Sirius, and slightly further away, at 11.5 light years. It shares with Sirius the peculiarity of having a white-dwarf partner, Procyon B, which is over 10,000 times fainter than Procyon and orbits it every 41 years. The two appear so close that only large telescopes, of the type used by professional astronomers, can separate them.

| Width: | Depth: | Area: 380 square degrees | Size ranking: 71st |

| Abbreviation: Cap | Genitive: Capricorni | Highest in sky at 10 pm: August to September |

Capricornus

The
Sea Goat

The smallest constellation of the zodiac, lying between Sagittarius and Aquarius, Capricornus depicts a fish-tailed goat. The constellation originated among the Babylonian and Sumerian peoples of the Middle East, but in Greek myth it is linked with the goat-like god Pan, who turned his lower half into a fish to escape the sea monster Typhon. Capricornus is not prominent: its brightest star, Delta (δ) Capricorni (Deneb Algedi), is of magnitude 2.8. The Sun passes through it from 19 January to 16 February.

**Fully visible
62°N–90°S**

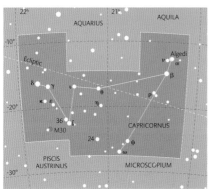

FEATURES OF INTEREST
Alpha (α) Capricorni (Algedi)
A wide double star. Ordinary binoculars, or good eyesight, reveal two stars, of magnitudes 3.6 and 4.3. Both are yellowish in colour, but they are unrelated. The brighter star is a giant, 110 light years away, while the fainter one is a supergiant, eight times more distant.

Beta (β) Capricorni A wide double star. The primary is a yellow giant, magnitude 3.1, visible to the naked eye. Binoculars or a small telescope show the second star, of magnitude 6.1. Both stars lie about 500 light years away.

| Width: | Depth: | Area: 414 square degrees | Size ranking: 40th |

Abbreviation: Car	Genitive: Carinae	Highest in sky at 10 pm: January to April

Carina

Carina is an impressive constellation, containing the second-brightest star of all, Canopus, and lying in a rich part of the Milky Way. In Greek times, it formed part of a much larger constellation, Argo Navis, representing the ship of the Argonauts, but was made separate in the 18th century by the Frenchman Nicolas Louis de Lacaille. It represents the ship's keel, with Canopus marking the rudder. The other parts of the ship – Vela (the sails) and Puppis (the stern) – lie to the north of Carina.

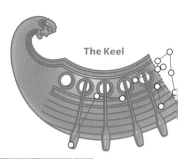

The Keel

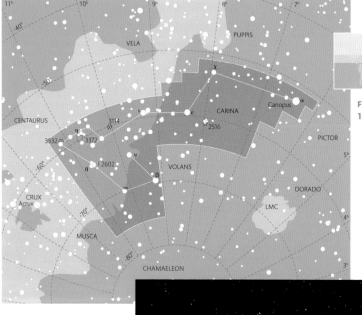

**Fully visible
14°N–90°S**

Star clusters and nebulae in Carina
This picture shows the whole of this magnificent southern constellation, with the bright star Canopus at the top right. The Eta Carinae Nebula (NGC 3372) lies on the far left, with the open cluster IC 2602 (also known as the Southern Pleiades) below and to its right. NGC 2516, another prominent cluster, is to the right of centre.

Width:	Depth:	Area: 494 square degrees	Size ranking: 34th

FEATURES OF INTEREST

Alpha (α) Carinae (Canopus) ◉ The second-brightest star in the sky, at magnitude -0.7. It is a white supergiant, 14,000 times more luminous than the Sun, lying 310 light years away.

Eta (η) Carinae ♠☆ A remarkable variable star. Currently it is of 6th magnitude, but it has been much brighter in the past, most noticeably in the 19th century, when it flared up to nearly -1. Eta Carinae is thought to be a supergiant with a mass of 100 Suns, one of the most massive stars known, orbited by an unseen companion. Through a telescope, it appears as a hazy orange ellipse, because of matter thrown off in its last outburst. Its distance is estimated at around 7,500 light years, and it lies within the extensive nebula NGC 3372 (see below).

NGC 2516 ◉♠ An open cluster, visible to the naked eye. Using binoculars, it is possible to pick out its individual stars, the brightest being a 5th-magnitude red giant, scattered over an area of sky the size of the full Moon. It is about 1,300 light years away.

NGC 3114 ◉♠ An open cluster, visible to the naked eye and appearing about the same size as the disc of the full Moon. Its individual stars can be seen through binoculars. It lies about 3,000 light years away.

IC 2602 (The Southern Pleiades)
This large, bright open cluster contains several stars visible to the naked eye. The bright star at the right of the picture is Theta (θ) Carinae.

NGC 3372 (The Eta Carinae Nebula) ◉♠☆
A large, bright diffuse nebula, four times the apparent width of the full Moon, surrounding the star Eta (η) Carinae (see above). The nebula is visible to the naked eye against the Milky Way and is best seen with binoculars. A V-shaped lane of dark dust runs through it. Near Eta Carinae itself, telescopes show a dark and bulbous cloud of dust called the Keyhole Nebula.

NGC 3532 ◉♠ A bright and dense open cluster,
elliptical in shape. It is visible to the naked eye and is a glorious sight through binoculars, being nearly twice the apparent diameter of the full Moon at its widest. It is 1,300 light years away.

IC 2602 (The Southern Pleiades) ◉♠
A large and prominent open cluster, containing eight stars brighter than magnitude 6. The brightest member is Theta (θ) Carinae, a blue-white star of magnitude 2.8. The cluster appears twice the size of the full Moon and lies about 500 light years away.

NGC 3372 (The Eta Carinae Nebula)
This magnificent nebula surrounds the peculiar variable star Eta (η) Carinae. It is dissected by dark lanes of dust. Eta Carinae itself is in the brightest part of the nebula, just above centre in this photograph.

Abbreviation: Cas	Genitive: Cassiopeiae	Highest in sky at 10 pm: October to December

Cassiopeia

This attractive constellation represents the mythical Queen Cassiopeia. Her husband and daughter are represented by the adjacent constellations Cepheus and Andromeda. Cassiopeia was notoriously vain, and is depicted sitting on a throne, fussing with her hair. Cassiopeia's brightest stars form a distinctive W-shape. Epsilon (ε) Cassiopeiae, at one end of the W, marks the queen's ankle while Beta (β), at the other end, lies in her shoulder.

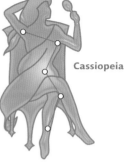

Cassiopeia

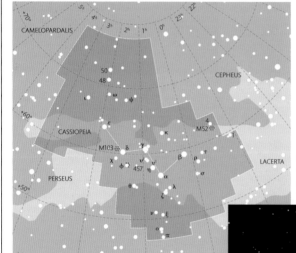

Fully visible
90°N–12°S

The main stars in Cassiopeia
The distinctive W-shape formed by Cassiopeia's brightest stars makes the constellation easy to recognize.

FEATURES OF INTEREST

Gamma (γ) Cassiopeiae ◉ A variable star, currently about magnitude 2.2. Its extremely rapid spin causes rings of gas to be thrown off its equator, changing its brightness temporarily.

Eta (η) Cassiopeiae ⚸ A double star. It consists of a yellow star of magnitude 3.5 and an orange companion of magnitude 7.5 that can be seen with a small telescope. The two stars lie 19 light years away, forming a genuine binary pair with an orbital period of 480 years.

Rho (ρ) Cassiopeiae ◉ ♙ A highly luminous yellow supergiant. As a result of pulsations in its size, it varies between about 4th and 6th magnitudes in a cycle that lasts about two years.

M52 ♙ ⚸ An open cluster. It is visible with binoculars, covering an area about one-third the apparent size

of the full Moon, although a telescope is needed to show its individual stars. It lies 5,200 light years away. A 5th-magnitude star that appears to be a member of the cluster actually lies much closer to us.

NGC 457 ♙ ⚸ An elongated open cluster, about one-third the apparent size of the full Moon and visible through binoculars or a small telescope. The cluster's appearance has been compared to an owl, with its two brightest stars marking the owl's eyes. The brightest star is the 5th-magnitude Phi (φ) Cassiopeiae, a luminous supergiant.

Width:	Depth:	Area: 598 square degrees	Size ranking: 25th

Abbreviation: Cen	Genitive: Centauri	Highest in sky at 10 pm: April to June

Centaurus

This large constellation in the southern Milky Way depicts a centaur, the mythical beast with the legs of a horse and the upper body of a man. In Greek mythology, this centaur is identified as Chiron, tutor to the offspring of the gods. Its brightest star, Alpha (α) Centauri, is the third brightest in the sky. A line drawn between Alpha (α) and Beta (β) Centauri points to the Southern Cross, Crux.

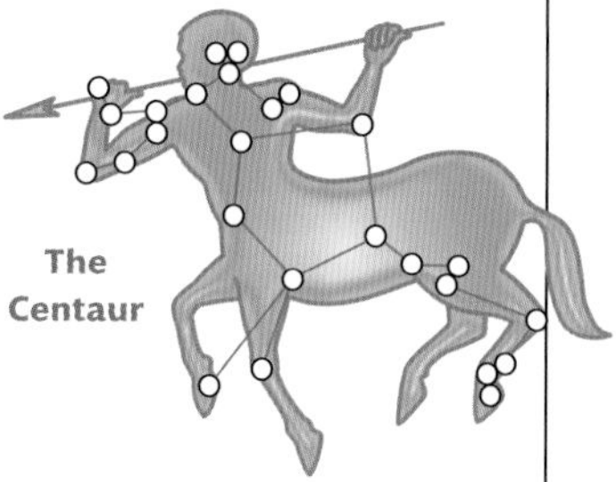

The Centaur

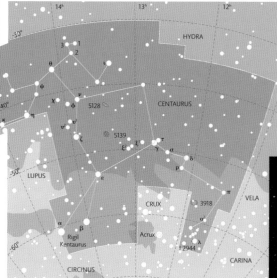

Fully visible
25°N–90°S

NGC 5139 (Omega Centauri)
When seen through a telescope, this bright globular cluster often appears elliptical rather than circular.

FEATURES OF INTEREST

Alpha (α) Centauri (Rigil Kentaurus) ◉ �
A nearby multiple star. To the naked eye, it appears of magnitude -0.3, the third-brightest star in the sky. A small telescope shows it to be a glorious binary, consisting of two yellow and orange stars of magnitudes 0.0 and 1.3. They orbit each other every 80 years, so one complete orbit can just be followed in a human lifetime. Lying 2 degrees away from this pair is a much fainter third star, an 11th-magnitude red dwarf called Proxima Centauri. This is the closest star to the Sun, 4.2 light years away, about 0.2 light years closer to us than the two brighter members of the Alpha Centauri system.

Beta (β) Centauri ◉ A blue-white giant star, magnitude 0.6 and 11th brightest in the sky, about 400 light years away.

NGC 5139 (Omega (ω) Centauri) ◉ ♠ ☀
The largest and brightest globular cluster in the sky, about 17,000 light years away. Seen with the naked eye or binoculars, it appears as a hazy 4th-magnitude patch larger than the full Moon. A small telescope shows its main stars.

NGC 3918 (The Blue Planetary) ☀
A planetary nebula, visible through a small telescope as a rounded blue disc.

NGC 5128 (Centaurus A) ☀ An unusual galaxy. With a small telescope, it looks like an elliptical galaxy, but larger apertures and long-exposure photographs show that it is bisected by a dark dust lane, evidently due to a merger with another galaxy. It is a strong radio source, referred to as Centaurus A.

Width:	Depth:	Area: 1,060 square degrees	Size ranking: 9th

Abbreviation: Cep	Genitive: Cephei	Highest in sky at 10 pm: September to October

Cepheus

This far-northern constellation, adjoining Cassiopeia and extending almost to the north celestial pole, represents the mythical King Cepheus, the husband of the vain Queen Cassiopeia and father of Andromeda. Its brightest star is Alpha (α) Cephei, of magnitude 2.5.

Cepheus

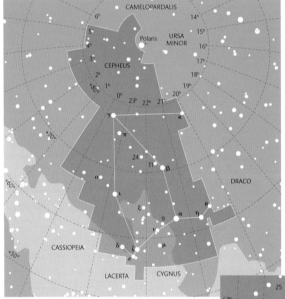

**Fully visible
90°N–1°S**

Delta and Mu Cephei

The following stars are useful for gauging the magnitude of Delta (δ) and Mu (μ) Cephei: Zeta (ζ), magnitude 3.4, Epsilon (ε), magnitude 4.2, and Lambda (λ), magnitude 5.1.

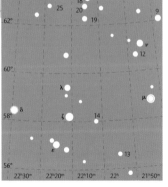

FEATURES OF INTEREST

Beta (β) Cephei ⚹ A double and variable star. The brighter component is a blue giant of magnitude 3.2, about 680 light years away. It is a pulsating variable star (the prototype of a group called Beta Cephei stars), although its variations are so small that they are barely perceptible to the eye. A small telescope shows an 8th-magnitude companion.

Delta (δ) Cephei ◉ ♁ ⚹ A double star and a famous variable. The brighter star, a yellow supergiant, is the prototype of the so-called Cepheid variables, which astronomers use for finding distances in space. Such stars change in brightness as they pulsate in size. Delta Cephei itself varies between magnitudes 3.5 and 4.4 every five days nine hours. It lies about 860 light years away. A small telescope shows a wide companion of magnitude 6.3.

Mu (μ) Cephei (The Garnet Star) ◉ ♁ ⚹
A variable star with a strong red colour, hence its popular name, which is noticeable through binoculars or a small telescope. It is a red supergiant that pulsates in size, varying from magnitude 3.4 to 5.1 about every two years.

Width: 🐫🐫	Depth: 🐫🐫🐫	Area: 588 square degrees	Size ranking: 27th

Abbreviation: Cet	Genitive: Ceti	Highest in sky at 10 pm: October to December

Cetus

Cetus, which straddles the celestial equator south of Pisces and Aries, is a large constellation representing the mythical sea monster from whose jaws Andromeda was rescued by the hero Perseus. The brightest star in the constellation is Beta (β) Ceti, magnitude 2.0.

The Sea Monster or Whale

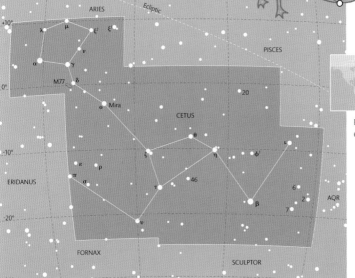

Fully visible 65°N–79°S

FEATURES OF INTEREST

Omicron (o) Ceti (Mira) 👁 ♏ A famous variable star. Mira is a red giant star, about 300 times larger than the Sun, that varies widely in brightness

Omicron (o) Ceti (Mira)
The variable star Mira (below centre in picture) is seen here at its maximum brightness.

as it swells and contracts in size in a cycle that lasts approximately 11 months. At its brightest, it can reach 3rd or even 2nd magnitude, but it drops to 10th magnitude at its minimum. Mira, which was first identified as being variable in 1596, is the prototype of a class of long-period variables; referred to as Mira stars, they form the largest group of variable stars. Mira is Latin for "wonderful".

Tau (τ) Ceti 👁 A nearby, Sun-like star. It is similar to the Sun in temperature and luminosity, and at a distance of 11.9 light years it is among the 20 closest stars to Earth. It appears of magnitude 3.5.

M77 ✶ A spiral galaxy, seen face-on from Earth. A telescope shows it as a small, round patch. It is the brightest of a class of galaxies known as Seyfert galaxies, which have particularly bright centres.

Width:	Depth:	Area: 1,231 square degrees	Size ranking: 4th

Abbreviation: Cha	Genitive: Chamaeleontis	Highest in sky at 10 pm: February to May

Chamaeleon

Chamaeleon is a small, faint constellation near the south celestial pole. Introduced at the end of the 16th century by the Dutch navigators Pieter Dirkszoon Keyser and Frederick de Houtman, it depicts a chameleon, the lizard that changes its skin color to camouflage itself. Its brightest stars are of 4th magnitude.

The Chameleon

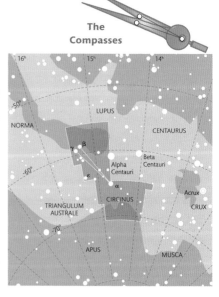

**Fully visible
7°N–90°S**

FEATURES OF INTEREST

Delta (δ) Chamaeleontis 🔭 A wide double star. With binoculars, it is easy to see the two components, of magnitudes 4.4 and 5.5. Both stars lie about 350 light years away.

NGC 3195 🔭 A faint planetary nebula, of similar apparent size to the disc of Jupiter, visible with a moderate-sized telescope.

Width:	Depth:	Area: 132 square degrees	Size ranking: 79th

Abbreviation: Cir	Genitive: Circini	Highest in sky at 10 pm: May to June

Circinus

This small constellation next to Centaurus was introduced in the 18th century by the French astronomer Nicolas Louis de Lacaille. It represents a pair of dividing compasses, of the type used by surveyors and chart makers, and is suitably placed in the sky next to Norma, the set square. Despite its insignificance, Circinus is easy to find, as it lies next to Alpha (α) Centauri in Centaurus. Although it is in the Milky Way, it contains no notable star clusters.

The Compasses

**Fully visible
19°N–90°S**

FEATURES OF INTEREST

Alpha (α) Circini 🔭 A double star. The primary, of magnitude 3.2, is the brightest star in the constellation. It lies 54 light years away. Its companion, of magnitude 8.6, is visible through a small telescope.

Width:	Depth:	Area: 93 square degrees	Size ranking: 85th

Abbreviation: Col	Genitive: Columbae	Highest in sky at 10 pm: January

Columba

The Dove

Columba was introduced in the late 16th century by the Dutch astronomer Petrus Plancius, who created it from stars near Canis Major. It represents Noah's dove.

FEATURES OF INTEREST
Alpha (α) Columbae ◉ A blue-white star, 260 light years away, of magnitude 2.6.

Mu (μ) Columbae ◉ ♙
A so-called runaway star, moving through our Galaxy at unusually high speed, about 100km (60 miles) per second. At least two fainter stars, 53 Arietis and AE Aurigae (not shown on the chart), seem to be diverging from the same point, south of Orion's belt. They might once have been part of a multiple star system that was disrupted when one of the stars exploded as a supernova, between two and three million years ago.

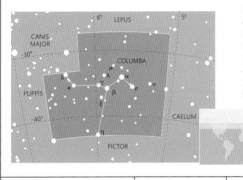

Fully visible 46°N–90°S

Width:	Depth:	Area: 270 square degrees	Size ranking: 54th

Abbreviation: Com	Genitive: Comae Berenices	Highest in sky at 10 pm: April to May

Coma Berenices

This constellation represents the hair of Queen Berenice of Egypt. According to legend, she cut off her locks to thank the gods for the safe return of her husband, King Ptolemy, from battle. It was formed in the 16th century by Caspar Vopel, a German cartographer, from a group of faint stars that the Greeks regarded as the tail of Leo. The galaxies in its southern region belong to the Virgo Cluster.

Berenice's Hair

NGC 4565
A telescope of moderate size reveals the shape of this spiral galaxy, seen edge-on. A lane of dark dust runs across it.

Fully visible 90°N–56°S

FEATURES OF INTEREST
M64 (The Black Eye Galaxy) ⚹ A spiral galaxy, visible through a small telescope as an elliptical haze. Larger apertures show a dark cloud of dust silhouetted against the galaxy's centre, which gives rise to its popular name.

The Coma Star Cluster (Melotte 111) ◉ ♙
An open cluster, consisting of a scattered group of stars stretching southwards in a fan shape from Gamma (γ) Comae Berenices, its brightest member at magnitude 4.4. The cluster lies 280 light years away.

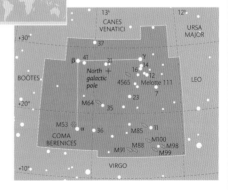

Width:	Depth:	Area: 386 square degrees	Size ranking: 42nd

Abbreviation: CrA	Genitive: Coronae Australis	Highest in sky at 10 pm: July to August

Corona Australis

The Southern Crown

The ancient Greeks visualized this small but pretty constellation as a wreath lying at the forefeet of Sagittarius. Its main feature is an arc of stars, distinctive even though the brightest is of only 4th magnitude.

FEATURES OF INTEREST

Gamma (γ) Coronae Australis ✶ A close binary, divisible through a telescope with a moderate-sized aperture. The stars, of 5th and 6th magnitude, orbit each other every 120 years. They lie 56 light years away.

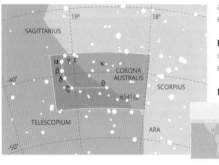

Kappa (κ) Coronae Australis ✶ A double star, with components of magnitudes 5.6 and 6.2, easily divided with a small telescope.

NGC 6541 ♙✶ A globular cluster, visible with a small telescope, or even binoculars. It covers an area about one-third the apparent size of the full Moon.

Fully visible 44°N–90°S

Width:	Depth:	Area: 128 square degrees	Size ranking: 80th

Abbreviation: CrB	Genitive: Coronae Borealis	Highest in sky at 10 pm: June

Corona Borealis

The Northern Crown

Corona Borealis is a semi-circular pattern lying between Boötes and Hercules. It depicts the crown worn by the mythical Princess Ariadne of Crete when she married the god Dionysus, who threw it into the sky, where its jewels turned into stars. The brightest of these stars, Alpha (α) Coronae Borealis, also called Alphecca, is of magnitude 2.2.

Arc of Corona Borealis
Seven stars form the Northern Crown.

FEATURES OF INTEREST

Zeta (ζ) Coronae Borealis ✶ A double star, with components of magnitudes 5.0 and 6.0, divisible with a small telescope.

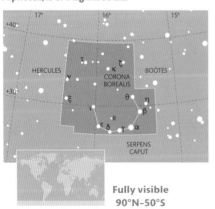

Nu (ν) Coronae Borealis ♙ A double star. Binoculars show the two stars, both red giants, of magnitudes 5.2 and 5.4; they lie about 650 and 540 light years away respectively.

R Coronae Borealis ♙✶ An unusual variable star, usually of around 6th magnitude. It is a highly luminous yellow supergiant about 4,000 light years away. Occasionally and unpredictably, it can fade within a few weeks to as faint as 15th magnitude, remaining there for some months before increasing in brightness again.

Fully visible 90°N–50°S

Width:	Depth:	Area: 179 square degrees	Size ranking: 73rd

| Abbreviation: Crv | Genitive: Corvi | Highest in sky at 10 pm: April to May |

Corvus

This small constellation south of Virgo represents a crow perched on the coils of Hydra, the water-snake. In Greek mythology, the crow was sent by Apollo to fetch water in a cup (represented by the adjoining constellation of Crater), but greedily stopped to eat figs instead. On its return, the crow blamed the water-snake for delaying it. But Apollo, who was not fooled by the lie, condemned the crow to a life of thirst, just out of reach of the cup in the heavens. Its brightest star, Gamma (γ) Corvi, is of magnitude 2.6.

The Crow

Fully visible 65°N–90°S

The main stars in Corvus
Four bright stars in Corvus form a distinctive shape called a keystone. Beta (β) Corvi is at the bottom left.

FEATURES OF INTEREST

Delta (δ) Corvi ⚹ A double star, with components of widely differing brightness. The brighter star, of magnitude 3.0, has a 9th-magnitude partner, which is visible through a small telescope.

NGC 4038 and 4039 (The Antennae) ⚹
A famous pair of interacting galaxies. They are small and of 11th magnitude, so a telescope of moderate to large aperture is needed to see them. Long-exposure photographs show two streamers of stars and gas that extend from the colliding galaxies like the feelers of an insect, giving rise to the popular name.

| Width: | Depth: | Area: 184 square degrees | Size ranking: 70th |

| Abbreviation: Crt | Genitive: Crateris | Highest in sky at 10 pm: April |

Crater

This undistinguished constellation represents a cup or chalice. In Greek myth, it is linked with its neighbouring constellations, Corvus, the crow, and Hydra, the water-snake. Crater's brightest star is Delta (δ) Crateris, magnitude 3.6, but it contains no objects of interest for observers using small telescopes.

The Cup

Fully visible 65°N–90°S

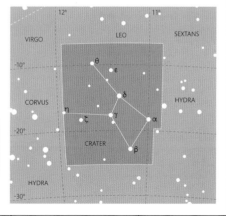

| Width: | Depth: | Area: 282 square degrees | Size ranking: 53rd |

Abbreviation: Cru	Genitive: Crucis	Highest in sky at 10 pm: April to May

Crux

The smallest constellation in the sky, Crux covers a mere five per cent of the area of the largest constellation, Hydra. Nevertheless, it is one of the most famous and easily recognized star patterns of all. To the ancient Greeks, its stars formed part of the hind legs of Centaurus, the centaur. Crux became a recognized constellation in its own right in the late 16th century as Europeans explored the southern oceans, although no one is credited with its invention. It lies in a brilliant region of the Milky Way. Its longer axis points to the south celestial pole.

The Southern Cross

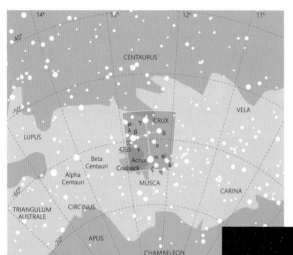

Fully visible 25°N–90°S

FEATURES OF INTEREST

Alpha (α) Crucis (Acrux) ◉ ♄ ⚹ A double star, appearing to the naked eye as a single object of magnitude 0.8, the 13th-brightest star in the sky. A small telescope divides it into a glittering blue-white pair, of magnitudes 1.3 and 1.8. Both stars are 320 light years away and probably form a true binary, although the orbital period is unknown. There is a much wider 5th-magnitude star, visible through binoculars, which is not related.

Beta (β) Crucis ◉ A blue-white giant, lying 280 light years away. At magnitude 1.3, it is among the 20 brightest stars in the sky. It pulsates in size five times a day, varying by under a tenth of a magnitude as it does so, not enough to be noticeable to the naked eye.

The Southern Cross
The four brightest stars of Crux form a distinctive shape known as the Southern Cross. A fifth bright star, Epsilon (ε) Crucis, lies close to the centre of the cross.

Width: ☝	Depth: ☝	Area: 68 square degrees	Size ranking: 88th

The Coalsack
This dark cloud of dust, seen below and to the left of the cross-shape of Crux, is easy to see with the naked eye. The stars that are visible in this area lie between us and the nebula.

NGC 4755 (The Jewel Box)
This sparkling open cluster was given its name in the 19th century by the English astronomer John Herschel, who described it as resembling a "casket of variously coloured precious stones".

Gamma (γ) Crucis ◉ ♄ A wide optical double star. The brighter star is a red giant of magnitude 1.6, 88 light years away. Binoculars show an unrelated 6th-magnitude companion, which is over three times more distant.

Iota (ι) Crucis ☌ A wide double star, with components of magnitudes 4.7 and 9.5 that can be separated with a small telescope.

Mu (μ) Crucis ◉ ☌ A wide optical double. A small telescope, or powerful binoculars, shows the two stars, of magnitudes 4.0 and 5.2. The fainter component is a rapidly rotating star that throws off rings of gas, causing occasional small variations in its brightness.

NGC 4755 (The Jewel Box, The Kappa (κ) Crucis cluster) ◉ ♄☌
A magnificent open cluster, visible to the naked eye as a bright knot in the Milky Way between Beta (β) Crucis and the Coalsack. It is, in fact, far more distant than either of them, lying 6,500 light years away. Binoculars or a small telescope show its individual stars, of 6th magnitude and fainter, covering an area of sky about one-third the apparent size of the full Moon. Most of its stars are blue-white supergiants, although near its centre there is a prominent red supergiant of 8th magnitude. The star near the centre

of the cluster that is actually designated Kappa (κ) Crucis is a blue-white supergiant of magnitude 5.9.

The Coalsack ◉ ♄ A prominent dark nebula, a wedge-shaped cloud of dust and gas about 600 light years away that blots out the light from the stars in the Milky Way behind it. The Coalsack is large enough for 13 full Moons to be lined up across it. It has no NGC or other catalogue number. Although the Coalsack lies mainly in Crux, it spills over the borders of Crux into Centaurus and Musca.

Using Crux to locate the south celestial pole
The south celestial pole lies roughly at the intersection of two imaginary lines. One is formed by extending the line between Alpha (α) and Gamma (γ) Crucis (top of picture); the other lies at right angles to a line joining Alpha (α) and Beta (β) Centauri (bottom).

Abbreviation: Cyg	Genitive: Cygni	Highest in sky at 10 pm: August to September

Cygnus

This large constellation depicts the swan into which, according to Greek myth, the god Zeus transformed himself for one of his illicit love trysts. One story says that the object of his desire was Queen Leda of Sparta, and that their union produced either one or two eggs, from which hatched the twins Castor and Pollux, as well as Helen of Troy. The swan's beak is marked by Beta (β) Cygni and its tail by Alpha (α) Cygni, named Deneb (from an Arabic word meaning "tail"). Deneb forms one corner of the Summer Triangle, with Vega (in Lyra) and Altair (in Aquila) marking the other two. Since it is also identifiable by its cross-shape, Cygnus is sometimes called the Northern Cross. It lies in a rich part of the Milky Way.

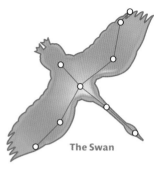

The Swan

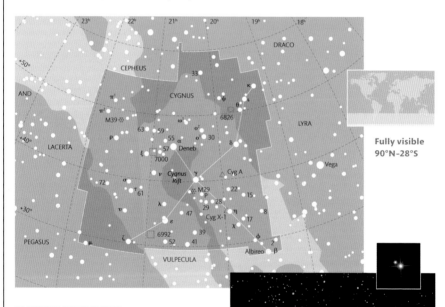

Fully visible
90°N–28°S

FEATURES OF INTEREST

Alpha (α) Cygni (Deneb) 👁 The brightest star in Cygnus, at magnitude 1.3. It is a luminous blue-white supergiant lying about 1,400 light years away, which makes it the most distant of all first-magnitude stars.

Beta (β) Cygni (Albireo) 🔭 A beautiful double star. The two stars, magnitudes 3.1 and 5.1, can be divided with powerful binoculars. They are easy to separate with a telescope, through which they present a striking contrast of orange for the brighter star and blue-green for the fainter one. Both stars lie about 400 light years away, but it is not known whether they form a true binary.

Cygnus (Inset: Albireo)
Cygnus, the swan, forms a large cross-shape in the northern Milky Way. The star marking the head of the swan (or the foot of the cross) is Albireo, a beautiful double star, divisible with a small telescope.

Width:	Depth:	Area: 804 square degrees	Size ranking: 16th

NGC 7000 (The North America Nebula)
This large nebula can be found between Deneb and Xi (ξ) Cygni. Its resemblance to the outline of the North American continent is evident in this photograph; a dark inlet similar in shape to the Gulf of Mexico can be made out at the bottom of the picture. The nebula is a bright cloud of gas, which is thought to be lit up by a partially obscured star within it.

Omicron-1 (o¹) Cygni (31 Cygni) ♐☀ An orange-coloured giant, magnitude 3.8, which has a wide 5th-magnitude, bluish companion (30 Cygni), ideal for binocular observation. A closer bluish star of 7th magnitude is also visible with binoculars or a small telescope.

Chi (χ) Cygni ◉♐ A pulsating red giant, one of the Mira class of variable stars (see p.13), that ranges in brightness between 3rd and 14th magnitudes every 13 months or so.

61 Cygni ☀ An attractive double star. The two components, of magnitudes 5.2 and 6.1, can be seen separately through a small telescope or even powerful binoculars. Both are orange dwarfs, smaller and fainter than the Sun, and are easily visible only because they are relatively close to us, 11.4 light years away. They orbit each other every 620 years.

M39 ♐☀ An open cluster, just visible to the naked eye under good conditions. Binoculars or a small telescope show its brightest stars. It covers the same area of sky as the full Moon, and lies about 1,000 light years away.

NGC 6826 (The Blinking Planetary) ☀ A planetary nebula visible with a small telescope as a bluish disc of similar size to the disc of Jupiter. Looking alternately at it and to one side gives the impression that it is blinking on and off, which explains its popular name.

NGC 6992 (The Veil Nebula) ♐☀ Part of a large, complex nebula called the Cygnus Loop which is the remains of a star that exploded as a supernova about 5,000 years ago. In clear skies, the Veil Nebula section can be seen with binoculars or a small telescope, but the whole nebula is best seen on photographs.

NGC 7000 (The North America Nebula) ♐☀
A large nebula, visible under dark skies with binoculars or a wide-field telescope, extending for up to four Moon diameters. It is best seen on long-exposure photographs, where it takes the shape of the continent of North America, hence its popular name.

Cygnus A ♔ A peculiar galaxy, thought to be two distant galaxies in collision. A large telescope is needed to see it, as it is of 15th magnitude. It is also a strong radio source.

Cygnus X-1 ♔ A possible black hole, and one of the strongest X-ray sources in the sky. It appears optically as a 9th-magnitude blue supergiant, about 8,000 light years away. The X-rays come from an invisible companion, which orbits the supergiant every 5.6 days. This companion is the supposed black hole.

NGC 6992 (The Veil Nebula)
The Veil Nebula, in the southern part of Cygnus, is the wispy remnant of an exploded star.

Abbreviation: Del	Genitive: Delphini	Highest in sky at 10 pm: August to September

Delphinus

This small but distinctive constellation is tucked in between Aquila and Pegasus. Delphinus is associated with two different Greek myths. According to one account, it represents the dolphin sent by Poseidon, the sea god, to fetch Amphitrite, a sea nymph, to be his bride. It has also been identified as the dolphin that saved Arion, a poet and musician, when he was attacked on board a ship by a gang of robbers. The constellation's two brightest stars are Alpha (α) and Beta (β) Delphini, magnitudes 3.8 and 3.6 respectively. These stars bear the unusual names Sualocin and Rotanev. When written backwards, the names spell Nicolaus Venator. This is the Latinized version of Niccolò Cacciatore, an astronomer at Palermo Observatory, Italy, who named the stars after himself in the early 19th century.

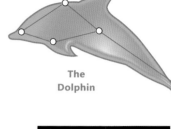

The Dolphin

The main stars in Delphinus
The attractive shape of Delphinus represents a leaping dolphin. Four stars, at the top left of this picture, form the shape called Job's Coffin.

**Fully visible
90°N–69°S**

FEATURES OF INTEREST

Job's Coffin ◉ ♙ A name given to the box shape formed by the four stars Alpha (α), Beta (β), Gamma (γ), and Delta (δ) Delphini, all of which are of 4th magnitude.

Gamma (γ) Delphini ✶ A double star. A small telescope will separate the two components, which appear yellow and white, of magnitudes 4.3 and 5.0. Both stars lie approximately 115 light years away. A fainter and closer pair of 8th-magnitude stars, known as Struve 2725 (about 120 light years away), should be visible in the same field of view.

Nova Delphini
This nova is now very faint but in 1967, when this photograph was taken, it flared to naked-eye visibility.

Width:	Depth:	Area: 189 square degrees	Size ranking: 69th

| Abbreviation: Dor | Genitive: Doradus | Highest in sky at 10 pm: December to January |

Dorado

Dorado, one of several southern constellations representing exotic creatures, was introduced in the late 16th century by the Dutch navigators Pieter Dirkszoon Keyser and Frederick de Houtman. It represents not the goldfish found in ornamental ponds but the dolphinfish of tropical seas, a member of the family Coryphaenidae. It has also been depicted as a swordfish. For astronomers, Dorado is significant because it contains the bulk of the Large Magellanic Cloud, our nearest neighbouring galaxy. Its brightest star is Alpha (α) Doradus, magnitude 3.3.

The "Goldfish" or Dolphinfish

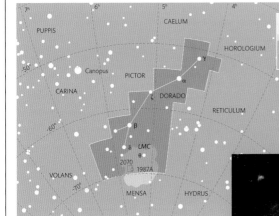

Fully visible 20°N–90°S

NGC 2070 (The Tarantula Nebula)
Loops of glowing gas extending from the Tarantula Nebula give it the appearance of a large spider.

FEATURES OF INTEREST

Beta (β) Doradus ◉ A bright Cepheid variable, ranging between magnitudes 3.5 and 4.1 every 9.8 days. It is a yellow-white supergiant and lies about 1,000 light years away.

NGC 2070 (The Tarantula Nebula) ◉ ♠ ✶ A bright nebula in the Large Magellanic Cloud (see right), visible to the naked eye. It appears about the same size as the full Moon, but its true diameter is about 800 light years, making it far larger than the Orion Nebula. The nebula is a star-forming region, and binoculars or a small telescope reveal a star cluster, 30 Doradus, at its centre.

Supernova 1987A ⚰ The brightest supernova visible from Earth since 1604. In 1987 it flared up in the Large Magellanic Cloud, near the Tarantula Nebula, reaching magnitude 2.8 at its brightest. It remained visible to the naked eye for 10 months.

The Large Magellanic Cloud (LMC) ◉ ♠ ✶ A small galaxy, a satellite of our own, lying 170,000 light years away. It resembles a detached part of the Milky Way, somewhat elongated in shape. It has about one-tenth the mass of our Galaxy and a true diameter of 20,000 light years. It is usually classified as an irregular galaxy, but there are traces of a barred spiral structure. Scanning across it with binoculars or a small telescope brings numerous star clusters and glowing nebulae into view. The largest and brightest nebula is NGC 2070, also known as the Tarantula Nebula (see above).

| Width: | Depth: | Area: 179 square degrees | Size ranking: 72nd |

Abbreviation: Dra	Genitive: Draconis	Highest in sky at 10 pm: March to September

Draco

This extensive constellation of the far-northern sky winds around Ursa Minor. It represents the dragon that in Greek mythology guarded the golden apples of the Hesperides, the daughters of Atlas, and which was slain by Hercules as one of his labours. In the sky, Hercules is represented with one foot on the dragon's head. Despite its considerable size, Draco contains no really prominent stars, the brightest being Gamma (γ) Draconis, magnitude 2.2, a red giant about 150 light years away. With the stars Beta (β), Nu (ν), and Xi (ξ) Draconis, it forms a lozenge shape that marks the head of the dragon. The constellation is noted chiefly for its double stars.

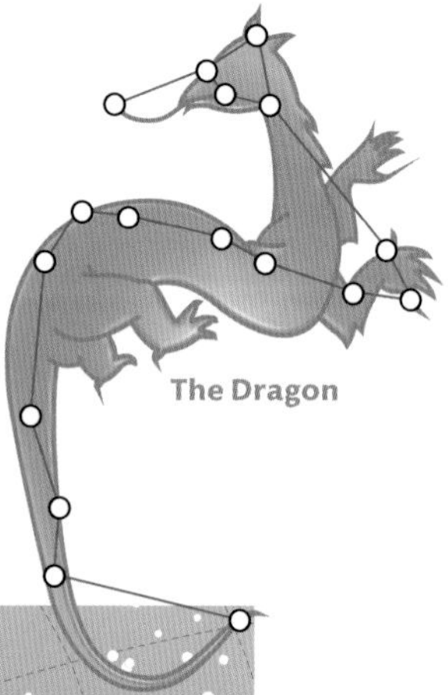

The Dragon

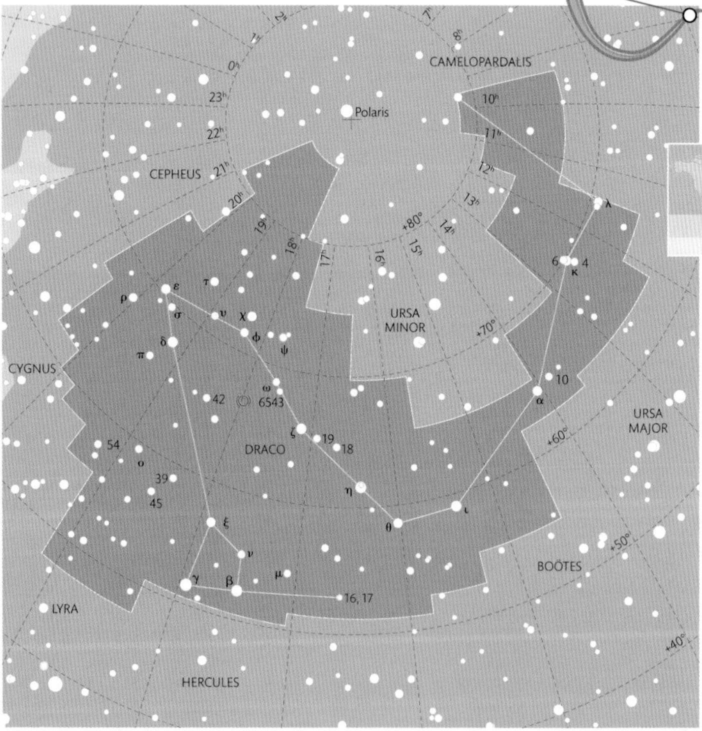

**Fully visible
90°N–4°S**

FEATURES OF INTEREST

Mu (μ) Draconis ✷ A double star, with components that are too close be separated with a small telescope. An aperture of at least 75mm (3in), with high magnification, will usually be needed to see the two 6th-magnitude stars. They are 88 light years away and orbit each other every 420 years.

Nu (ν) Draconis ♙✷ A wide double star with twin white 5th-magnitude components that are virtually identical in colour and brightness. The pair form an excellent sight through binoculars and are easy to see with the smallest telescopes. Both stars lie 100 light years away.

Width: 🖐🖐🖐	Depth: 🖐🖐	Area: 1,083 square degrees	Size ranking: 8th

Psi (ψ) Draconis 🔭 A double star, easy to divide through a small telescope. The two components, of magnitudes 4.6 and 5.8, may even be divided with powerful binoculars.

16 and 17 Draconis 🔭 A multiple star. Binoculars show a wide double with 5th- and 6th-magnitude components. A telescope with high magnification reveals that the brighter star, 17 Draconis, is itself a close double.

39 Draconis 🔭 A multiple star. Through a small telescope, or even binoculars, it appears as a wide pair of 5th- and 8th-magnitude stars. Examination of the brighter star under high magnification will show a much closer star, also of 8th magnitude.

NGC 6543 🔭 A planetary nebula, visible with a small telescope. It appears as a bluish disc of similar size to the globe of Saturn.

The head of Draco
A group of four stars represents the head of Draco. The brightest of them, at the lower left in this photograph, is Gamma (γ) Draconis.

Abbreviation: Equ	Genitive: Equulei	Highest in sky at 10 pm: September

Equuleus

The second-smallest constellation in the sky, Equuleus represents a small horse that lies next to Pegasus, the well-known flying horse. It first appeared among the 48 constellations described by the Greek astronomer Ptolemy, but there appear to be no legends about it. Its brightest star is Alpha (α) Equulei, magnitude 3.9.

The
Little Horse
or Foal

Fully visible
90°N–77°S

FEATURES OF INTEREST
Gamma (γ) Equulei 🔭 An optical double star, divisible with binoculars. It consists of a star of magnitude 4.7 with a very wide companion, 6 Equulei, of magnitude 6.1.

Epsilon (ε) Equulei 🔭 A multiple star. A small telescope shows it as a double, with components of magnitudes 5.4 and 7.4. The brighter star is, in fact, a binary with an orbital period of 100 years, but the stars are too close together to be separated with a small aperture.

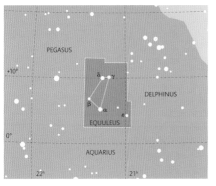

Width:	Depth:	Area: 72 square degrees	Size ranking: 87th

| Abbreviation: Eri | Genitive: Eridani | Highest in sky at 10 pm: November to January |

Eridanus

This long, straggling constellation represents a river in Greek mythology. The River Eridanus features in the story of Phaethon, who fell into it after a disastrous attempt to drive the chariot of his father, the sun-god Helios. The constellation extends for nearly 60 degrees from north to south, the greatest range in declination of any constellation.

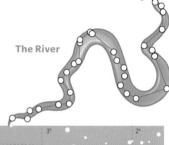

The River

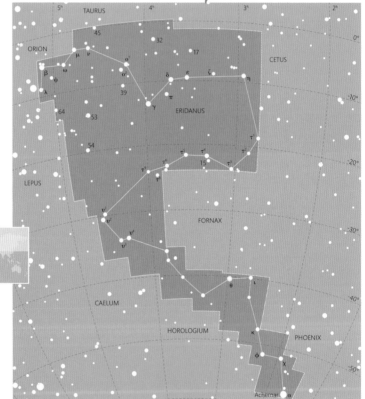

Fully visible
32°N–89°S

Omicron-2 (o^2) Eridani

All three components of this multiple star are visible in this telescopic view. The primary is in the centre with the two dwarf stars, overlapping each other, to its right.

FEATURES OF INTEREST

Alpha (α) Eridani (Achernar) ◉ A blue-white star of magnitude 0.5, the brightest star in Eridanus and the ninth brightest in the sky. It lies 140 light years away. Its name, Achernar, is of Arabic origin, meaning "river's end", and the star marks the constellation's southernmost tip. Nearby is p Eridani, a wide pair of 6th-magnitude stars visible with a small telescope.

| Width: | Depth: | Area: 1,138 square degrees | Size ranking: 6th |

Epsilon (ε) Eridani ◉ One of the closest naked-eye stars to the Sun. It is 10.5 light years distant and appears of magnitude 3.7. Although somewhat cooler and fainter than the Sun, it is otherwise similar.

Theta (θ) Eridani ⚹ An attractive double star. The two components, twin white stars of magnitudes 3.2 and 4.1, can be separated with a small telescope.

Omicron-2 (o²) Eridani (40 Eridani) ⚹ A multiple star with three components, including a red dwarf and a white dwarf together. The brightest of the trio, magnitude 4.4, is similar to the Sun. A small telescope shows that this star has a 10th-magnitude companion, which is the white dwarf; this is the easiest white dwarf to see with small instruments. It forms a binary with an 11th-magnitude red dwarf, although a larger aperture may be needed to spot this. The remarkable trio lies 16.5 light years away.

32 Eridani ⚹ A double star, with components of contrasting colours. A small telescope reveals a red giant of magnitude 4.7 and a blue-white companion of magnitude 6.1.

The southern tip of Eridanus
A small part of this large constellation can be seen here. Achernar is just above the horizon, while Chi (χ) and Phi (φ) Eridani are above and to the left of it.

Abbreviation: For	Genitive: Fornacis	Highest in sky at 10 pm: November to December

Fornax

Fornax is an undistinguished constellation in an unremarkable area of sky. It lies on the shores of the celestial river Eridanus and the southern border of Cetus. Invented in the 18th century by the French astronomer Nicolas Louis de Lacaille, it represents a chemical furnace.

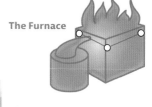

The Furnace

Fully visible
50°N–90°S

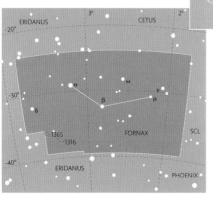

FEATURES OF INTEREST
Alpha (α) Fornacis ⚹ The brightest star in the constellation, appearing to the naked eye as a single star of magnitude 3.8. A small telescope shows that it has a 7th-magnitude companion. The two stars form a true binary, orbiting each other every 270 years or so. They lie 46 light years away from the Earth.

The Fornax Cluster ⚹ A cluster of galaxies about 60 million light years away, the brightest of which, notably NGC 1316 and NGC 1365, can be seen with a small telescope. NGC 1316 is a peculiar spiral, also known as radio source Fornax A. NGC 1365 is a barred spiral.

Width:	Depth:	Area: 398 square degrees	Size ranking: 41st

| Abbreviation: Gem | Genitive: Geminorum | Highest in sky at 10 pm: January to February |

Gemini

A constellation of the zodiac, Gemini depicts the mythological twins Castor and Pollux, after whom its two brightest stars are named. The twins sailed with the Argonauts in search of the golden fleece, and they were later regarded by the ancient Greeks as patron saints of seafarers. The two stars themselves are not related, though, lying at different distances from us. Gemini sits between Taurus and Cancer, and the Sun passes through it from 21 June to 20 July.

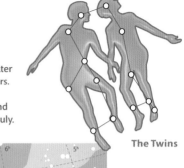

The Twins

Fully visible 90°N–55°S

FEATURES OF INTEREST

Alpha (α) Geminorum (Castor) ◉ ➤

A remarkable multiple star. To the naked eye, it appears as a single star, of magnitude 1.6. A small telescope divides it into a blue-white pair of stars, of magnitudes 1.9 and 3.0. These form a genuine binary with an orbital period of about 450 years. Both of these stars are spectroscopic binaries. A wider companion, of 9th magnitude, can also be seen with a small telescope. This is, in fact, a close pair of red dwarfs, forming an eclipsing binary. The whole six-star family is just over 50 light years away.

Castor and Pollux

The two brightest stars in Gemini are Castor (right) and Pollux (left). Although it has been assigned the letter Beta (β), Pollux is actually the brighter star.

| Width: | Depth: | Area: 514 square degrees | Size ranking: 30th |

Beta (β) Geminorum (Pollux) ◉ � The brightest star in the constellation and among the 20 brightest in the sky, at magnitude 1.1. It is an orange-coloured giant, 34 light years away. The coloration is more noticeable when the star is viewed through binoculars.

Zeta (ζ) Geminorum ◉ � A Cepheid variable star, ranging between magnitudes 3.6 and 4.2 in a cycle lasting just over 10 days. It is a yellow supergiant, about 1,400 light years away. Binoculars show a wide 8th-magnitude companion, which is unrelated.

Eta (η) Geminorum ◉ � A variable red giant, about 700 light years away. It pulsates in size in a cycle lasting about 8 months, varying between magnitudes 3.2 and 3.9 as it does so.

M35 �⚹ A rich open cluster, just visible to the naked eye and easy to see with binoculars. It appears almost as large as the full Moon. Binoculars or a small telescope resolve its individual stars, of 8th magnitude and fainter. The cluster lies nearly 3,000 light years away.

NGC 2392 (The Clown-Face Nebula, The Eskimo Nebula) ⚹ A planetary nebula. Its bluish disc, similar in size to the globe of Saturn, is visible through a small telescope, but large apertures are needed to detect the surrounding features that lend it the appearance of a face and give rise to its popular names.

M35
The open cluster M35 at the centre of this image contains about 200 stars. Inspection with a telescope reveals that the stars are arranged in curved lines.

Abbreviation: Gru	Genitive: Gruis	Highest in sky at 10 pm: September to October

Grus

Grus is one of the dozen southern constellations introduced at the end of the 16th century by the Dutch navigators Pieter Dirkszoon Keyser and Frederick de Houtman. It represents the long-necked wading bird of the family Gruidae. Its brightest star is Alpha (α) Gruis, known as Alnair, magnitude 1.7.

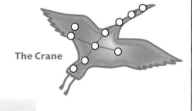

The Crane

Fully visible 33°N–90°S

FEATURES OF INTEREST

Beta (β) Gruis ◉ A variable red giant, 177 light years away, that fluctuates from magnitude 2.0 to 2.3 with no set period.

Delta (δ) Gruis ◉ A double star, with components that can be separated with the naked eye. Both stars lie about 300 light years away. Delta-1 is a yellow giant of magnitude 4.0; Delta-2 is a red giant of magnitude 4.1.

Mu (μ) Gruis ◉ � A wide naked-eye double. Both stars are yellow giants: Mu-1 (μ¹) is of magnitude 4.8 and Mu-2 (μ²) is 5.1. Both stars lie about 240 light years away.

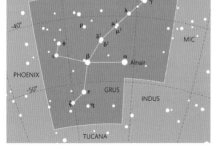

Width: 👓👓	Depth: 👓👓	Area: 366 square degrees	Size ranking: 45th

Abbreviation: Her	Genitive: Herculis	Highest in sky at 10 pm: June to August

Hercules

A large but not prominent constellation depicting the hero of Greek myth, Hercules lies between the bright stars Arcturus and Vega. The body of Hercules is inverted in the sky, the head being marked by Alpha (α) Herculis, in the south, and the feet by the stars to the north. Hercules was ordered by King Eurystheus of Mycenae to perform 12 labours, one of which was to slay a dragon (marked by adjacent Draco). Hercules is depicted resting on his right knee, with his left foot on the dragon's head. The constellation features the brightest globular cluster in northern skies, M13, and some notable double stars. Its brightest star is Beta (β) Herculis, magnitude 2.8.

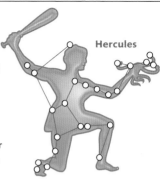

Hercules

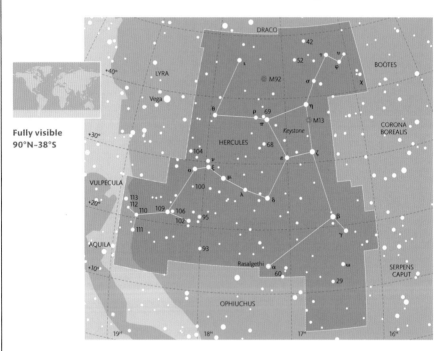

Fully visible 90°N–38°S

FEATURES OF INTEREST

Alpha (α) Herculis (Rasalgethi) ◉ ✴ A red giant of variable brightness. It ranges between 3rd and 4th magnitudes, with no set period, as a result of fluctuations in its size. A small telescope shows a companion of magnitude 5.3. Hercules is depicted kneeling, and the name Rasalgethi is derived from an Arabic term meaning "the kneeler's head".

Rho (ρ) Herculis ✴ An optical double star. The two components, of magnitudes 4.6 and 5.4, can be separated using a small telescope with high magnification.

95 Herculis ✴ A silver and gold pair of stars, of magnitudes 5.0 and 5.2, divisible through a small telescope.

100 Herculis ✴ A pair of almost identical white stars, both of magnitude 5.8, that can be separated with a small telescope. The two stars do not form a true binary.

Width:	Depth:	Area: 1,225 square degrees	Size ranking: 5th

M13 ◉ ♙ ⚹ The brightest globular cluster in the northern sky. Under dark skies, it appears to the naked eye like a hazy star. Binoculars show it clearly, about half the apparent width of the full Moon, and a small telescope shows its brightest stars. It is 25,000 light years away.

M92 ♙ ⚹ A globular cluster, fainter and smaller than M13 but still worthy of attention. Binoculars are needed to find it, and a telescope of moderate aperture will show its stars.

Hercules, including the Keystone
Pi (π), Eta (η), Zeta (ζ), and Epsilon (ε) Herculis (at the centre of this picture) form a four-sided figure called the Keystone that marks the pelvis of Hercules.

M13
This spectacular cluster lies on one side of the Keystone, a third of the way between Eta (η) and Zeta (ζ) Herculis. It contains hundreds of thousands of stars.

Abbreviation: Hor	Genitive: Horologii	Highest in sky at 10 pm: November to December

Horologium

This unrenowned southern constellation, invented in the 18th century by the French astronomer Nicolas Louis de Lacaille, lies between the bright stars Achernar and Canopus. It represents a pendulum clock, with its brightest star, Alpha (α) Horologii (magnitude 3.9), marking the pendulum weight.

The Pendulum Clock

FEATURES OF INTEREST
R Horologii ♙ ⚹ A variable red giant, of the type known as a Mira star. It ranges between 5th and 14th magnitudes every 13 months or so.

TW Horologii ♙ A variable red giant. Over a period of approximately five or six months, it ranges between about 5th and 6th magnitudes.

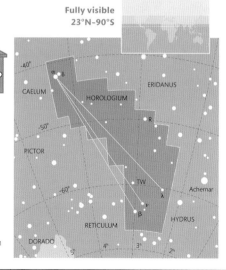

Fully visible
23°N–90°S

Width:	Depth:	Area: 249 square degrees	Size ranking: 58th

Abbreviation: Hya	Genitive: Hydrae	Highest in sky at 10 pm: February to June

Hydra

Despite being the largest of all the 88 constellations, Hydra is far from prominent. Its head, formed by a loop of six stars of 3rd and 4th magnitudes, lies just north of the celestial equator, under Cancer, while the tip of its tail is 90 degrees away, between Libra and Centaurus. In Greek mythology, Hydra was a multi-headed monster slain by Hercules as one of his 12 labours. During the fight, Hercules was attacked by a crab, represented by the constellation Cancer. Hydra is also linked with a moral tale involving a crow sent to fetch a cup of water; the crow and cup are depicted by the constellations Corvus and Crater, which are represented lying on Hydra's back.

The Water-snake

Fully visible
54°N–83°S

M83
The bright core and spiral arms of this galaxy, seen face-on from Earth, are visible on photographs and digital images, such as this one taken using a CCD camera.

FEATURES OF INTEREST

Alpha (α) Hydrae (Alphard) ◉ An orange-coloured giant about 180 light years away. The name Alphard comes from an Arabic term that means "the solitary one". It is an apt name because, at magnitude 2.0, Alphard is the only star in Hydra brighter than 3rd magnitude.

Epsilon (ε) Hydrae ⚹ A double star. A telescope of moderate aperture is needed to split the two close components, of magnitudes 3.4 and 6.5, which orbit each other every 400 years or so. They lie 130 light years away.

R Hydrae ◉ ♠ ⚹ A variable red giant of the Mira type that ranges between 3rd and 11th magnitudes every 13 months or so.

U Hydrae ♠ A variable red giant that fluctuates between 4th and 6th magnitudes with little regularity

as the star changes in size. It lies about 680 light years away.

M48 ♠ ⚹ An open cluster, just visible to the naked eye under good conditions but best seen with binoculars or a wide-field telescope. It is larger in apparent size than the full Moon and lies about 2,500 light years away.

M83 ⚹ A spiral galaxy that appears face-on to us, so that its arms are clearly visible through a large-aperture telescope and on photographs. When viewed through a small telescope, it appears as a rounded patch.

NGC 3242 (The Ghost of Jupiter) ⚹
A planetary nebula, visible through a small telescope as a bluish, planet-like disc, hence its popular name.

Width:	Depth:	Area: 1,303 square degrees	Size ranking: 1st

| Abbreviation: Hyi | Genitive: Hydri | Highest in sky at 10 pm: October to December |

Hydrus

This constellation of the far-southern skies was invented in the late 16th century by the Dutch navigators Pieter Dirkszoon Keyser and Frederick de Houtman. Lying between the bright star Achernar and the south celestial pole, it depicts a small water-snake. It should not be confused with the large water-snake, Hydra, known since the time of the ancient Greeks. The constellation's brightest star is Beta (β) Hydri, magnitude 2.8.

The Little Water-snake

Fully visible 8°N–90°S

FEATURES OF INTEREST

Pi (π) Hydri A wide pair of unrelated stars, easy to divide through binoculars. Both are red giants. Pi-1 (π¹) is of magnitude 5.6 and lies about 700 light years away; Pi-2 (π²), magnitude 5.7, is about 490 light years away.

| Width: | Depth: | Area: 243 square degrees | Size ranking: 61st |

| Abbreviation: Ind | Genitive: Indi | Highest in sky at 10 pm: August to October |

Indus

The Indian

This southern constellation, lying between Pavo and Tucana, was introduced in the late 16th century by the Dutch navigators Pieter Dirkszoon Keyser and Frederick de Houtman. It depicts a native Indian, although from exactly where remains unclear; on old maps, the figure is shown holding arrows and a spear. Indus's brightest star, Alpha (α) Indi, is of magnitude 3.1.

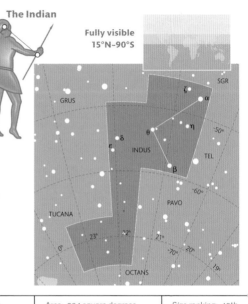

Fully visible
15°N–90°S

FEATURES OF INTEREST

Epsilon (ε) Indi ◉ Among the nearest stars to us, Epsilon Indi is only 11.9 light years away and has a magnitude of 4.7. It is a somewhat smaller and cooler star than the Sun.

Theta (θ) Indi ⟩ A double star, consisting of components of magnitudes 4.5 and 6.9 that are visible through a small telescope.

| Width: | Depth: | Area: 294 square degrees | Size ranking: 49th |

| Abbreviation: Lac | Genitive: Lacertae | Highest in sky at 10 pm: September to October |

Lacerta

The Lizard

Lacerta is a small and insignificant constellation of the northern sky on the edge of the Milky Way, sandwiched between Cygnus and Andromeda. It was introduced in the 17th century by the Polish astronomer Johannes Hevelius. Its brightest star is Alpha (α) Lacertae, magnitude 3.8. Three naked-eye novae erupted within its boundaries during the 20th century.

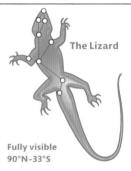

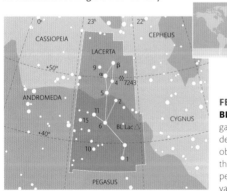

Fully visible
90°N–33°S

FEATURES OF INTEREST

BL Lacertae ✦ The nucleus of a distant elliptical galaxy, originally classified as a variable star, hence its designation. It has given its name to a class of similar objects, the BL Lacertae (or BL Lac) objects, which are thought to be galaxies with powerful energy sources, perhaps massive black holes, at their centres. BL Lacertae varies between about 12th and 16th magnitudes.

| Width: | Depth: | Area: 201 square degrees | Size ranking: 68th |

Abbreviation: Leo	Genitive: Leonis	Highest in sky at 10 pm: March to April

Leo

This large, impressive constellation depicts a crouching lion. In Greek myth, this was the lion that Hercules killed as one of his 12 labours. Leo is a constellation of the zodiac, lying between Cancer and Virgo. The Sun passes through it from 10 August to 16 September.

The Lion

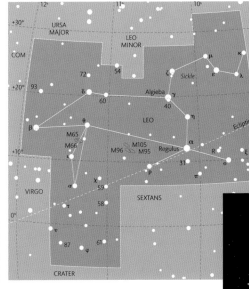

**Fully visible
82°N–57°S**

The main stars in Leo
Leo's main stars form a shape that closely resembles the outline of a crouching lion. Regulus and the Sickle are at the right of the picture.

FEATURES OF INTEREST

The Sickle ◉ An easily recognizable pattern of six stars, shaped like a reversed question mark or a hook, which forms the head and chest of the lion. It consists of the stars Epsilon (ε), Mu (μ), Zeta (ζ), Gamma (γ), Eta (η), and Alpha (α) Leonis.

Alpha (α) Leonis (Regulus) ◉♄⚹ The faintest of the first-magnitude stars, magnitude 1.4. It is a blue-white star about 79 light years away. Binoculars or a small telescope show that it has a wide companion of 8th magnitude.

Gamma (γ) Leonis (Algieba) ⚹ A glorious double star, consisting of two golden orange giants of magnitudes 2.4 and 3.6, which make an excellent sight through a small telescope. They are a genuine binary pair, orbiting each other every 500 years or so. Binoculars (or even sharp eyesight) show a much wider 5th-magnitude star, 40 Leonis, which is unrelated.

Zeta (ζ) Leonis ♄ A wide triple, formed by unrelated stars. Zeta itself is of magnitude 3.4. Binoculars show a 6th-magnitude star, 35 Leonis, to its north. Further south is 39 Leonis, also of 6th magnitude.

R Leonis ♄⚹ A red-giant Mira variable, ranging in brightness between about 4th and 11th magnitudes every 10 months or so.

M65 and M66 ⚹ Two 9th-magnitude spiral galaxies, visible with a small telescope. Being tilted at an angle to us, they appear elliptical.

M95 and M96 ⚹ A pair of spiral galaxies, visible through a small telescope as elongated smudges. They lie at similar distances to M65 and M66, namely 30–35 million light years.

Width: 🖐🖐🖐	Depth: 🖐🖐🖐	Area: 947 square degrees	Size ranking: 12th

| Abbreviation: LMi | Genitive: Leonis Minoris | Highest in sky at 10 pm: March to April |

Leo Minor

This faint constellation, representing a lion cub, is squeezed
between Leo and Ursa Major. It was invented in the 17th century
by Johannes Hevelius. Its brightest star is 46 Leonis Minoris,
magnitude 3.8, but it contains little to interest owners of
small instruments.

The
Lesser Lion

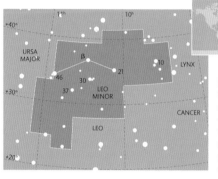

**Fully visible
90°N–48°S**

FEATURES OF INTEREST

Beta (β) Leonis Minoris ♛ A close double star. It
appears to the naked eye as a single star of magnitude
4.2, and its components can be separated only
through a telescope of very large aperture. The two
stars, which are 150 light years away, have an orbital
period of 38 years. This is the only star in the
constellation to have been assigned a Greek letter.

| Width: 🖐🖐 | Depth: ✋ | Area: 232 square degrees | Size ranking: 64th |

| Abbreviation: Lep | Genitive: Leporis | Highest in sky at 10 pm: January |

Lepus

Lepus lies under the feet of Orion, the hunter, and
is pursued across the sky by his dog Canis Major.
The constellation was known to the ancient Greeks.
Its brightest star is Alpha (α) Leporis, magnitude
2.6, whose name, Arneb, comes
from an Arabic term meaning
"the hare".

The Hare

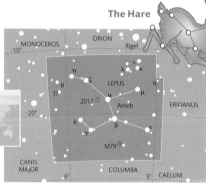

**Fully visible
62°N–90°S**

FEATURES OF INTEREST

Gamma (γ) Leporis 🔭 A double star, with
components of magnitudes 3.6 and 6.2 that
are visible through binoculars. Both stars are at
a similar distance from us, about 30 light years.

R Leporis 🔭⭐ A Mira variable, noted for its
deep red colour. It ranges from about 6th to
12th magnitude every 14 months or so.

M79 ⭐ A small globular cluster of 8th magnitude,
visible with a small telescope. It lies about 40,000
light years away. In the same field of view is a triple
star called h3752 (the h stands for John Herschel,

who first catalogued it). This consists of a close pair
of stars, of 5th and 7th magnitudes, with a wider
companion of 9th magnitude.

NGC 2017 ⭐ A small star cluster, the main members
of which are catalogued as the multiple star h3780.
A small telescope reveals a 6th-magnitude star with
four companions of 8th to 10th magnitude. A larger
aperture will separate two of the components into
close doubles, and there is also a fainter 12th-
magnitude star that completes the group.

| Width: ✋ | Depth: ✋ | Area: 290 square degrees | Size ranking: 51st |

Abbreviation: Lib	Genitive: Librae	Highest in sky at 10 pm: May to June

Libra

A constellation of the zodiac, Libra lies just south of the celestial equator, between Virgo and Scorpius. Originally it was seen to represent the claws of Scorpius, the scorpion, which is why its brightest stars have names that mean the northern and southern claw. Libra was first visualized as a balance by the Romans, over 2,000 years ago, and it is now usually depicted as the scales held by the adjacent figure of Virgo, who is seen as the goddess of justice. The Sun passes through Libra from 31 October to 23 November.

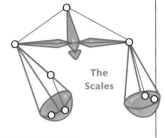

The Scales

**Fully visible
60°N–90°S**

The main stars in Libra
Alpha (α) Librae is at the centre right of this picture, with Beta (β) above and to its left.

FEATURES OF INTEREST

Alpha (α) Librae (The Southern Claw, Zubenelgenubi) 👁 🄼 A wide double star. The two components, of magnitudes 2.8 and 5.2, can be separated with binoculars or even good eyesight.

Beta (β) Librae (The Northern Claw, Zubeneschamali) 🄼 🔭 The constellation's brightest star, at magnitude 2.6. It is said by some observers to have a greenish tinge, which is highly unusual. Binoculars or a small telescope should reveal its colour.

Delta (δ) Librae 🄼 An eclipsing binary of the same type as Algol (in Perseus). Its variations in magnitude, from 4.9 to 5.9 over 2 days 8 hours, are easy to follow through binoculars.

Iota (ι) Librae 🄼 🔭 A complex multiple star, appearing to the naked eye as a single star of magnitude 4.5. Binoculars show a wide 6th-magnitude companion, 25 Librae. A telescope with an aperture of 75mm (3in) or more shows that the primary has a 9th-magnitude partner. High magnification should reveal that this fainter companion is itself a close pair.

Mu (μ) Librae 🔭 A double star, with components of magnitudes 5.6 and 6.6. A telescope with an aperture of about 75mm (3in) is needed to separate them.

Width: 👐	Depth: 👐	Area: 538 square degrees	Size ranking: 29th

Abbreviation: Lup	Genitive: Lupi	Highest in sky at 10 pm: May to June

Lupus

Lupus lies in the Milky Way to the south of Libra. It represents a wolf that in Greek and Roman times was visualized as being held on a pole by a centaur, depicted by the adjacent Centaurus. It seems, however, that there are no myths specifically about the wolf. Lupus's brightest star is Alpha (α) Lupi, magnitude 2.3.

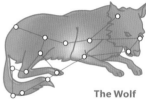

The Wolf

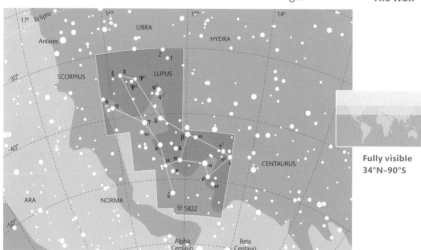

**Fully visible
34°N–90°S**

FEATURES OF INTEREST

Kappa (κ) Lupi ⚹ A double star. The components, of magnitudes 3.8 and 5.5, are easy to divide with a small telescope.

Mu (μ) Lupi ⚹ A multiple star. With a small telescope, it appears as a wide double star with components of magnitudes 4.3 and 6.8. An aperture of 100mm (4in) or more, with high magnification, shows that the primary is, in fact, a close pair of 5th-magnitude stars.

Xi (ξ) Lupi ⚹ A double star. Its components, of magnitudes 5.1 and 5.6, can be separated with a small telescope.

Pi (π) Lupi ⚹ A matching pair of 5th-magnitude blue-white stars. A telescope with an aperture of 75mm (3in) or more is needed to see them separately.

NGC 5822 ♉⚹ A rich open cluster, of similar apparent size to the full Moon. Visible with binoculars or a small telescope, it contains over 100 faint stars and lies 2,700 light years away.

NGC 5822
This large open cluster is shown here as it would look through a small telescope.

Width: 🤲	Depth: 🤲	Area: 334 square degrees	Size ranking: 46th

Abbreviation: Lyn	Genitive: Lyncis	Highest in sky at 10 pm: January to March

Lynx

The Polish astronomer Johannes Hevelius, who invented this faint constellation in the 17th century, named it the Lynx because, he said, only the lynx-eyed would be able to spot it. It lies in the northern sky between Ursa Major and Auriga, and is surprisingly large – greater in area than Gemini, for example. Except in good dark-sky conditions, naked-eye observers will see little more than its brightest star, Alpha (α) Lyncis, magnitude 3.1. There are, however, numerous double stars to attract telescope users.

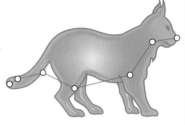

The Lynx

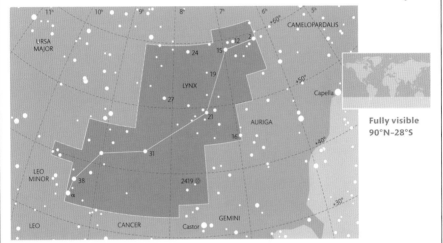

Fully visible 90°N–28°S

FEATURES OF INTEREST

12 Lyncis ⚹ A multiple star. Through a small telescope, it appears double, with components of magnitudes 4.9 and 7.1. An aperture of 75mm (3in) or more reveals that the brighter star is a close pair of 5th- and 6th-magnitude stars, which orbit each other every 700 years.

19 Lyncis ⚹ An easily divided multiple star. A small telescope separates it into a double, with components of magnitudes 5.8 and 6.7. Further away, a third star, of 8th magnitude, should also be visible.

38 Lyncis ⚹ A tight double star, requiring a telescope of 75mm (3in) aperture to split it into components of magnitudes 3.9 and 6.1.

NGC 2419 ⬙ A globular cluster, notable for its remoteness. At 300,000 light years away, it is more distant than the Magellanic Clouds and so appears small and of only 10th magnitude.

NGC 2419
It is thought likely that this cluster, seen here in a CCD image, lies outside our Galaxy in intergalactic space.

Width: 🖐🖐	Depth: 🖐🖐	Area: 545 square degrees	Size ranking: 28th

Abbreviation: Lyr	Genitive: Lyrae	Highest in sky at 10 pm: July to August

Lyra

This prominent constellation of the northern sky lies between Cygnus and Hercules. It represents the lyre played by Orpheus, the great musician of Greek mythology. Arab astronomers, though, visualized the pattern as an eagle, and the name of its brightest star, Vega, comes from their term meaning "swooping eagle". Vega forms one corner of a large triangle of stars known to northern observers as the Summer Triangle, and completed by Deneb (in Cygnus) and Altair (in Aquila).

The Lyre

Fully visible
90°N–42°S

Delta (δ) Lyrae 🔭 A wide pair of unrelated stars, divisible with binoculars or the naked eye. One is a red giant that varies slightly, between about magnitudes 4.2 and 4.3, and the other is a blue-white star of magnitude 5.6.

Epsilon (ε) Lyrae (The Double Double) 🔭🔭 A striking quadruple star. When viewed with binoculars, or even sharp eyesight, it appears as a pair of 5th-magnitude stars. A telescope with an aperture of 60–75mm (2½–3in), with high magnification, will divide each star into a closer binary. The slightly wider pair, of magnitudes 5.0 and 6.1, has a calculated orbital period of about 1,750 years. The other pair, of magnitudes 5.2 and 5.4, has an orbital period of over 700 years. All four stars lie about 160 light years away.

FEATURES OF INTEREST

Alpha (α) Lyrae (Vega) 👁 The fifth-brightest star in the sky, at magnitude 0.0. It is a blue-white star, 25 light years away.

Beta (β) Lyrae 🔭 A double star, the brighter component of which is variable. Beta Lyrae itself is an eclipsing binary that varies between magnitudes 3.3 and 4.4 in a cycle that lasts 12 days 22 hours. A small telescope shows that it has a wide companion of magnitude 7.1.

Zeta (ζ) Lyrae 🔭🔭 A double star. The two components, magnitudes 4.4 and 5.6, are easy to divide with binoculars or a small telescope.

M57 (The Ring Nebula) 🔭 A planetary nebula, visible through a small telescope as an elliptical disc. Larger apertures show it as a ring, which is how it appears on photographs.

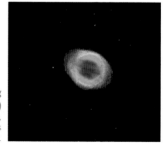

Epsilon (ε) Lyrae and Vega
The famous Double Double, Epsilon (ε) Lyrae, is at the top left in this picture, with the bright star Vega to its right.

M57 (The Ring Nebula)
Through a small telescope, this planetary nebula is visible as an elliptical disc.

Width: ✋	Depth: 👀	Area: 286 square degrees	Size ranking: 52nd

Microscopium | 109

| Abbreviation: Men | Genitive: Mensae | Highest in sky at 10 pm: December to February |

Mensa

This small and faint constellation near the south celestial pole was introduced in the 18th century by the French astronomer Nicolas Louis de Lacaille. He named it after Table Mountain, near the modern Cape Town in South Africa, from where he charted the southern sky. Mensa's only notable feature is part of the Large Magellanic Cloud, which extends into it from neighbouring Dorado.

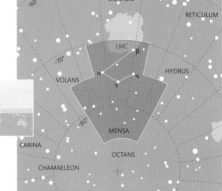

The Table Mountain

Fully visible 5°N–90°S

FEATURES OF INTEREST
Alpha (α) Mensae ◉ ♏ A yellow star, 33 light years away and similar in nature to the Sun. At magnitude 5.1, it is the brightest star in the constellation.

| Width: | Depth: | Area: 153 square degrees | Size ranking: 75th |

| Abbreviation: Mic | Genitive: Microscopii | Highest in sky at 10 pm: August to September |

Microscopium

Microscopium is a faint constellation to the south of Capricornus. It was introduced in the 18th century by the French astronomer Nicolas Louis de Lacaille and represents a microscope. The constellation's brightest stars, Gamma (γ) and Epsilon (ε) Microscopii, are both of magnitude 4.7.

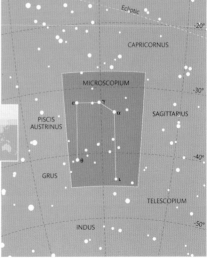
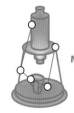

The Microscope

Fully visible 45°N–90°S

FEATURES OF INTEREST
Alpha (α) Microscopii ✶ An orange giant star of magnitude 4.9. A telescope shows a companion of 10th magnitude.

| Width: | Depth: | Area: 210 square degrees | Size ranking: 66th |

| Abbreviation: Mon | Genitive: Monocerotis | Highest in sky at 10 pm: January to February |

Monoceros

Monoceros straddles the celestial equator between Orion and Canis Minor. Introduced in the early 17th century by the Dutchman Petrus Plancius, it represents the mythical unicorn. Its brightest star is Alpha (α) Monocerotis, magnitude 3.9. Monoceros is often overlooked in favour of its glittering neighbouring constellations, but it lies in the Milky Way and contains much of interest for owners of any size of instrument.

The Unicorn

Fully visible 78°N–78°S

FEATURES OF INTEREST

Beta (β) Monocerotis ✸ A triple star, divisible with a small telescope. The three stars, magnitudes 4.6, 5.0, and 5.3, form an arc.

8 Monocerotis (Epsilon (ε) Monocerotis) ✸ A double star, with components of magnitudes 4.4 and 6.6 that are easy to divide through a small telescope.

S Monocerotis (15 Monocerotis) ✸ A highly luminous blue-white star, magnitude 4.7 (but slightly variable), in the cluster NGC 2264 (see next page). It has a companion of 8th magnitude, visible with a small telescope.

M50 ♍✸ An open cluster, about half the apparent size of the full Moon and visible with binoculars. A small telescope reveals its individual stars, of 8th magnitude and fainter. It lies 3,000 light years away.

NGC 2244 and the Rosette Nebula
The stars of the open cluster NGC 2244 (at the centre of this picture) are enveloped in the Rosette Nebula.

| Width: | Depth: | Area: 482 square degrees | Size ranking: 35th |

NGC 2232 ♦ A scattered open cluster. The same apparent size as the full Moon, it is just visible to the naked eye. Its brightest star is 5th-magnitude 10 Monocerotis, and several other individual member stars are visible with binoculars. It lies 1,100 light years away.

NGC 2244 ♦ An open cluster at the heart of the much larger Rosette Nebula. The cluster is easily visible through binoculars as an elongated group about two-thirds the apparent width of the full Moon. However, excellent skies are needed to trace the outline of the surrounding Rosette Nebula, which is three to four times larger and shows up well only on photographs. The cluster and nebula lie about 5,000 light years away.

NGC 2264 ♦ An open cluster. It is visible with binoculars, and when viewed through a small telescope it appears triangular in shape. Its brightest member is S Monocerotis (see previous page). Long-exposure photographs show a surrounding area of faint nebulosity, which includes a dark lane known as the Cone Nebula. The cluster and associated nebula are about 2,400 light years away.

The Cone Nebula
The Cone Nebula is a tapering column of dark dust in the southern part of the faint area of nebulosity that surrounds the star cluster NGC 2264.

| Abbreviation: Mus | Genitive: Muscae | Highest in sky at 10 pm: April to May |

Musca

Musca lies in the Milky Way, to the south of Centaurus and Crux. It depicts a fly and is one of the dozen southern constellations invented in the late 16th century by the Dutch navigators Pieter Dirkszoon Keyser and Frederick de Houtman. Its brightest star is Alpha (α) Muscae, magnitude 2.7.

The Fly

Fully visible
14°N–90°S

FEATURES OF INTEREST
Beta (β) Muscae ⊁ A close binary, which appears through a small telescope as a single star of magnitude 3.0. An aperture of around 100mm (4in) will divide it into two stars of magnitudes 3.6 and 4.0, which orbit each other every 400 years or so.

Theta (θ) Muscae ⊁ A double star. The two components, of magnitudes 5.6 and 7.6, are divisible with a small telescope. The brighter star is a luminous blue supergiant, while its companion is an example of a Wolf-Rayet star, a hot star that has lost its outer layers.

| Width: | Depth: | Area: 138 square degrees | Size ranking: 77th |

| Abbreviation: Nor | Genitive: Normae | Highest in sky at 10 pm: June |

Norma

Norma lies in the Milky Way between Ara and Lupus. Introduced in the 18th century by the French astronomer Nicolas Louis de Lacaille, it depicts a draughtsman's set square. It adjoins another Lacaille invention, Circinus, the compasses. There are no stars labelled Alpha (α) or Beta (β) Normae due to boundary changes since Lacaille's time.

The Set Square

Fully visible 29°N–90°S

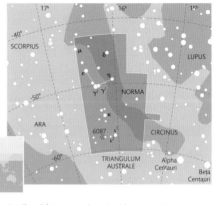

FEATURES OF INTEREST

Gamma (γ) Normae ◉ A double star, consisting of unrelated components that can be separated with the naked eye. The primary, Gamma-2 (γ²), is the brightest star in the constellation; it is an orange giant star of magnitude 4.0, 130 light years away. Gamma-1 (γ¹), magnitude 5.1, is an immensely luminous yellow supergiant more than 100 times further away.

Epsilon (ε) Normae ✶ A double star, with components of magnitudes 4.5 and 6.1 that can be separated with a small telescope.

Iota-1 (ι¹) Normae ✶ A double star. The components, of magnitudes 4.6 and 8.0, can be divided with a small telescope. Iota-2 (ι²) is an unrelated 6th-magnitude star that lies some distance away.

| Width: | Depth: | Area: 165 square degrees | Size ranking: 74th |

| Abbreviation: Oct | Genitive: Octantis | Highest in sky at 10 pm: October |

Octans

Octans contains the south celestial pole. The constellation, which represents a navigator's octant, a predecessor of the sextant, was introduced in the 18th century by the French astronomer Nicolas Louis de Lacaille. Its brightest star is Nu (ν) Octantis, an orange giant of magnitude 3.7. Other than its location at the south celestial pole, there is little notable about Octans.

Fully visible 0°–90°S

The Octant

FEATURES OF INTEREST

Sigma (σ) Octantis ◉ This is the nearest naked-eye star to the south celestial pole (lying about 1 degree away), although at magnitude 5.4 it is hardly prominent. It is a yellow-white giant star, 290 light years away.

Lambda (λ) Octantis ✶ A close double star, divisible with a small telescope. The two components are of magnitudes 5.5 and 7.2.

| Width: | Depth: | Area: 291 square degrees | Size ranking: 50th |

| Abbreviation: Oph | Genitive: Ophiuchi | Highest in sky at 10 pm: June to July |

Ophiuchus

The Serpent Holder

This large constellation extends from Hercules in the north, across the celestial equator, to Scorpius in the south. It represents the Greek god of medicine, Asclepius, who is depicted holding a serpent (the constellation Serpens), a traditional symbol for healing. The constellation contains several globular clusters. Its brightest star is Alpha (α) Ophiuchi, magnitude 2.1, also known as Rasalhague, derived from an Arabic term meaning "head of the serpent bearer".

Fully visible 59°N–75°S

M10
This is the most impressive of several globular clusters in Ophiuchus.

FEATURES OF INTEREST

Rho (ρ) Ophiuchi A multiple star. Binoculars show it as a star of magnitude 4.6 with two wide companions, of magnitudes 6.8 and 7.3. A small telescope, with high magnification, reveals that the brightest star has a closer companion of magnitude 5.7.

36 Ophiuchi A pair of near-identical orange dwarf stars of 5th magnitude, divisible with a small telescope. They lie 20 light years away and orbit each other every 500 years or so.

70 Ophiuchi A binary star, consisting of yellow and orange dwarfs, magnitudes 4.2 and 6.1, that lie 17 light years away and orbit each other every 88 years. They are slowly moving apart as seen from Earth, and should currently be divisible with all but the smallest apertures.

Barnard's Star The second-closest star to the Sun, 5.9 light years away. It is a cool and faint red dwarf of magnitude 9.5, and is hence far too faint to be seen with the naked eye.

M10 and M12 Two globular clusters about 3 degrees apart. Both are visible through binoculars, appearing about half the size of the full Moon, although a telescope of moderate aperture is needed to resolve individual stars. M10 is the closer cluster at 14,000 light years, compared with 16,000 light years for M12.

NGC 6633 An open cluster of similar apparent size to the full Moon, visible through binoculars.

IC 4665 A large and scattered open cluster, visible with binoculars.

| Width: | Depth: | Area: 948 square degrees | Size ranking: 11th |

| Abbreviation: Ori | Genitive: Orionis | Highest in sky at 10 pm: December to January |

Orion

Orion is the most magnificent of all the constellations. Being positioned on the celestial equator, it is visible from most places on Earth. It represents a hunter with his dogs (marked by Canis Major and Canis Minor) at his heels. In Greek mythology, Orion was the son of Poseidon, the sea god. He was supposedly killed by the sting of a scorpion, and his position in the sky is such that he sets as the scorpion (the constellation Scorpius) rises. According to another story, Orion became enamoured of a group of nymphs called the Pleiades, depicted by a star cluster in the adjoining constellation of Taurus. As the Earth turns, Orion seems to chase the Pleiades across the sky. The constellation contains several bright stars, but its most celebrated feature is the huge nebula (M42) that lies within the hunter's sword, south of the line of three stars marking his belt.

Orion

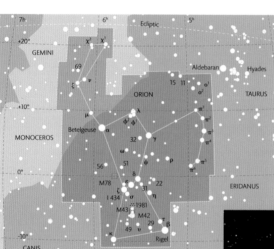

Fully visible
79°N–67°S

Orion
The magnificent constellation of Orion, with Betelgeuse at the top left, Rigel at the bottom right, and the Orion Nebula and the stars of the belt in the centre.

FEATURES OF INTEREST

Alpha (α) Orionis (Betelgeuse) ◉ A red supergiant star of variable brightness, ranging from about magnitude 0 to 1.6 every 6 years or so. It lies 500 light years away. The name Betelgeuse is derived from an Arabic term that incorporates a reference to a hand, although the star actually lies on the hunter's shoulder.

Beta (β) Orionis (Rigel) ◉ A blue supergiant star of magnitude 0.1, the brightest star in the constellation and the seventh brightest in the sky. It is 860 light years away. The name Rigel is derived from an Arabic word meaning "foot", which is the part of the hunter's body that the star represents.

| Width: 🖐🖐 | Depth: 🖐🖐 | Area: 594 square degrees | Size ranking: 26th |

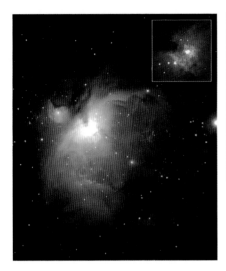

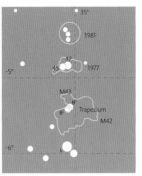

Detail of the M42 (Orion Nebula) region

Magnitude key

●	1
●	2
●	3
●	4
●	5
●	6
●	7

Theta-1 (θ¹) Orionis (The Trapezium) ✶ A multiple star, located at the heart of the Orion Nebula (see below). A small telescope shows it as a quadruple star, with components of magnitudes 5.1, 6.7, 6.7, and 8.0 arranged in a trapezium shape. A larger aperture reveals two other 11th-magnitude stars in the group.

Theta-2 (θ²) Orionis ♠ A double star. The two stars, magnitudes 5.0 and 6.4, are divisible through binoculars. Theta-2 also forms a wide, bright binocular double with Theta-1 Orionis.

Iota (ι) Orionis ♠ ✶ A double star. The two components, of magnitudes 2.8 and 7.0, can be split with a small telescope. Binoculars show another double nearby: called Struve 747, it consists of stars of magnitudes 4.7 and 5.5.

Sigma (σ) Orionis ✶ A remarkable multiple star. The main star, of magnitude 3.8, has two 7th-magnitude companions to one side, and a 9th-magnitude companion on the other. A faint triple star, called Struve 761, should be visible in the same telescopic field of view.

M42 (The Orion Nebula) ◉ ♠ ✶ One of the most spectacular objects in the sky. It is a cloud of glowing gas with an apparent diameter over twice that of the full Moon. Visible to the naked eye, it becomes larger and more complex when viewed with binoculars and telescopes of increasing aperture. It is 1,500 light years away and is lit up by the stars of the Trapezium (see Theta-1 (θ¹) Orionis, above) that lie within it. A northern extension of the nebula is known as M43.

M42 (Orion Nebula) with The Trapezium inset
M42 is a mass of glowing gas where stars are forming. At its heart is the multiple star Theta-1 (θ¹) Orionis.

NGC 1981 ♠ A large, scattered open cluster visible with binoculars, its brightest stars being of 6th magnitude. The cluster appears to the north of the Orion Nebula and lies at almost the same distance from us, 1,300 light years.

The Horsehead Nebula ♟ A dark nebula, shaped like a chess knight, seen silhouetted against a strip of brighter nebulosity that extends south from Zeta (ζ) Orionis. However, it is too faint to be viewed without a large telescope, and its shape is most easily seen on a long-exposure photograph.

The Horsehead Nebula
This distinctively shaped cloud of gas is visible against a background of brighter hydrogen gas.

| Abbreviation: Pav | Genitive: Pavonis | Highest in sky at 10 pm: July to September |

Pavo

Pavo, which represents a peacock, is one of the 12 southern constellations invented in the late 16th century by the Dutch navigators Pieter Dirkszoon Keyser and Frederick de Houtman. The constellation's brightest star, Alpha (α) Pavonis, magnitude 1.9, is also named Peacock. Being a modern constellation, there are no myths associated with it.

The Peacock

Fully visible
15°N–90°S

FEATURES OF INTEREST

Kappa (κ) Pavonis ◑ A Cepheid variable star. It is a yellow-white supergiant that ranges between magnitudes 3.9 and 4.8 every 9 days 2 hours. It lies about 630 light years away.

Xi (ξ) Pavonis ⚹ A close double star with components of unequal brightness. It consists of a red giant of magnitude 4.4 and an 8th-magnitude companion, both about 440 light years away. With a telescope of small aperture, it may be difficult to pick out the fainter star in the glare around the brighter component.

NGC 6752 ♁⚹ A globular cluster, just visible to the naked eye and easy to see through binoculars or a small telescope. A telescope with an aperture of 75mm (3in) will resolve individual stars. The cluster has an apparent size about half that of the full Moon and lies 14,000 light years away.

NGC 6752
This large globular cluster, one of the brightest in the sky, has many outlying stars.

| Width: | Depth: | Area: 378 square degrees | Size ranking: 44th |

| Abbreviation: Peg | Genitive: Pegasi | Highest in sky at 10 pm: September to October |

Pegasus

The Winged Horse

This large constellation north of Aquarius and Pisces adjoins Andromeda. It represents the upper body of the winged horse that, in Greek mythology, sprang from the body of Medusa when she was beheaded by Perseus. The most distinctive feature in Pegasus is the Great Square formed by Alpha (α), Beta (β), and Gamma (γ) Pegasi, and Alpha (α) Andromedae. The area inside the square is relatively barren, containing no stars brighter than 4th magnitude.

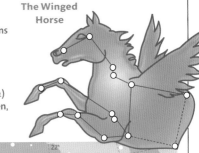

Fully visible 90°N–53°S

M15
This is one of the finest globular clusters in northern skies, appearing through binoculars like a hazy star.

FEATURES OF INTEREST

Beta (β) Pegasi ◉ A variable red giant that forms one corner of the Great Square. It varies unpredictably from magnitude 2.3 to 2.7.

Epsilon (ε) Pegasi (Enif) ♉⚹ A wide double star, which consists of a yellow supergiant of magnitude 2.4 and an 8th-magnitude partner that can be seen with a small telescope or even good binoculars. This star represents Pegasus's muzzle; its popular name is derived from the Arabic word for "nose".

M15 ♉⚹ A globular cluster, clearly visible with binoculars or a small telescope and about one-third the apparent width of the full Moon. It lies about 30,000 light years away.

| Width: | Depth: | Area: 1,121 square degrees | Size ranking: 7th |

| Abbreviation: Per | Genitive: Persei | Highest in sky at 10 pm: November to December |

Perseus

Perseus represents the mythological Greek hero who decapitated the fearsome Medusa, whose gaze could turn men to stone. On his way back from this exploit, Perseus rescued Andromeda from the jaws of a sea monster. In the sky, Perseus lies next to Andromeda and her mother, Cassiopeia, forming part of a great tableau depicting this most famous of Greek myths. Perseus is represented brandishing his sword in his right hand, marked by the twin star clusters NGC 869 and NGC 884, while in his left hand he holds the head of Medusa, marked by the star Beta (β) Persei, better known as Algol. A rich part of the Milky Way runs through Perseus, making it an attractive constellation for binocular users.

Perseus

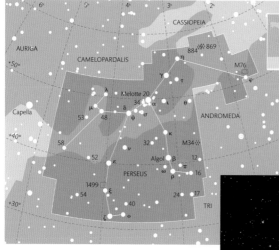

Fully visible
90°N–31°S

NGC 869 and NGC 884 (The Double Cluster)
These twin star clusters both lie about 8,000 light years away, in the Perseus spiral arm of the Galaxy. NGC 869 is the richer of the two.

FEATURES OF INTEREST

Alpha (α) Persei A yellow-white supergiant of magnitude 1.8, the brightest star in the constellation. It is the most prominent member of a large and loose cluster of stars known as Melotte 20, visible with binoculars, which lies about 600 light years away.

Beta (β) Persei (Algol) A famous eclipsing binary star, consisting of a close pair of stars in orbit around each other. When the brighter star is eclipsed by the fainter one, every 2 days 21 hours, the magnitude drops from 2.1 to 3.4 for about 10 hours.

Rho (ρ) Persei A variable red giant. Due to changes in size, it varies from magnitude 3.3 to 4.0 in a cycle lasting about 7 weeks.

M34 An open cluster, about 1,600 light years away. It is visible with binoculars or a small telescope and appears about the same size as the full Moon. The cluster's brightest stars are of 7th magnitude.

NGC 869 and NGC 884 (h and χ Persei, The Double Cluster) Two open clusters, just visible to the naked eye and an excellent sight through binoculars or a small telescope. Each cluster covers about the same area of sky as the full Moon.

| Width: | Depth: | Area: 615 square degrees | Size ranking: 24th |

Abbreviation: Phe	Genitive: Phoenicis	Highest in sky at 10 pm: October to November

Phoenix

Phoenix lies near the southern end of Eridanus, close to the bright star Achernar. It is the largest of the 12 constellations invented at the end of the 16th century by the Dutch navigators Pieter Dirkszoon Keyser and Frederick de Houtman, and represents the mythical bird that was supposedly reborn from the dead body of its predecessor. Its brightest star is Alpha (α) Phoenicis, magnitude 2.4.

The Phoenix

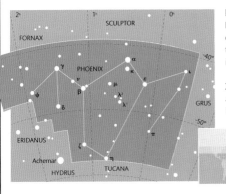

FEATURES OF INTEREST
Beta (β) Phoenicis ⚹ A double star. The two stars, of magnitudes 4.0 and 4.1, are not easy to divide: a telescope with an aperture of at least 100mm (4in) is needed to separate them.

Zeta (ζ) Phoenicis ◑♨⚹ A star that is both variable and double. The brighter component, an eclipsing binary of the same type as Algol, ranges between magnitudes 3.9 and 4.4 every 1 day 16 hours. The fainter star, of 8th magnitude, can be seen with a small telescope.

Fully visible 32°N–90°S

Width:	Depth:	Area: 469 square degrees	Size ranking: 37th

Abbreviation: Pic	Genitive: Pictoris	Highest in sky at 10 pm: December to February

Pictor

Pictor is one of the constellations representing instruments of science and the arts that were introduced in the 18th century by Nicolas Louis de Lacaille. It depicts an easel, and lies due south of Columba between the bright star Canopus in Carina and the Large Magellanic Cloud. Its brightest star is Alpha (α) Pictoris, magnitude 3.3.

The Painter's Easel

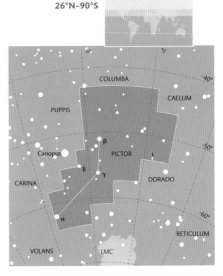

Fully visible 26°N–90°S

FEATURES OF INTEREST
Beta (β) Pictoris ▮ A blue-white star of magnitude 3.9, lying 63 light years away. Astronomers have found that it is surrounded by a disc of dust, thought to be a planetary system in the early stages of formation.

Iota (ι) Pictoris ⚹ A double star. The two components, of magnitudes 5.6 and 6.2, are easy to separate with a small telescope.

Width:	Depth:	Area: 247 square degrees	Size ranking: 59th

| Abbreviation: Psc | Genitive: Piscium | Highest in sky at 10 pm: October to November |

Pisces

This constellation of the zodiac, lying between Aquarius and Aries, depicts two fishes whose tails are each tied with cord. The star Alpha (α) Piscium marks a knot that joins the two cords. The constellation originated among the Babylonians of the Middle East, from whom it was inherited by the ancient Greeks. In one Greek myth, the fish represent Aphrodite and her son Eros, who plunged into the Euphrates to escape Typhon, a multi-headed monster. The Sun lies in Pisces from 12 March to 18 April and hence is in the constellation at the March equinox.

The Fishes

**Fully visible
83°N–56°S**

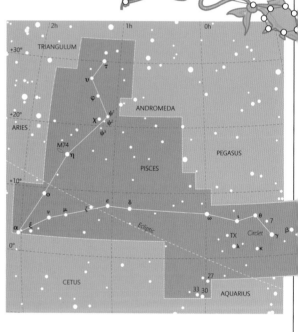

M74
The structure of this spiral galaxy, which appears face-on to us, can be seen through a large telescope.

FEATURES OF INTEREST

The Circlet ◉ A ring of seven stars of 4th and 5th magnitudes, south of the Great Square of Pegasus, that represents the body of the more southerly of the two fish. It consists of Gamma (γ), Kappa (κ), Lambda (λ), TX (or 19), Iota (ι), Theta (θ), and 7 Piscium.

Alpha (α) Piscium ⚹ A binary star. A telescope with an aperture of 75mm (3in) or more is needed to separate the close blue-white components of magnitudes 4.1 and 5.1. The two stars, which lie 160 light years away, have an orbital period of about 2,000 years and are currently moving closer together as seen from Earth.

Zeta (ζ) Piscium ⚹ A wide double star. The components, of magnitudes 5.2 and 6.3, both lie 130 light years away and can be divided with a small telescope.

Psi-1 (ψ¹) Piscium ⚹ A wide double star, with components of magnitudes 5.3 and 5.5 that can be separated with a small telescope. Both stars lie about 280 light years away.

TX Piscium (19 Piscium) ◉ ♈ A variable red giant that ranges from about magnitude 4.8 to 5.2 with no set period.

M74 ⚹ A spiral galaxy, visible through a small telescope as a rounded misty patch.

| Width: | Depth: | Area: 889 square degrees | Size ranking: 14th |

Abbreviation: PsA	Genitive: Piscis Austrini	Highest in sky at 10 pm: September to October

Piscis Austrinus

Piscis Austrinus is a small constellation of the southern sky that was known to the ancient Greeks. It depicts a fish into whose mouth Aquarius, represented by the constellation to the north, pours water from his urn. In Greek mythology, this fish was also the parent of the two fish represented by Pisces. In the past, the constellation was also known as Piscis Australis.

The Southern Fish

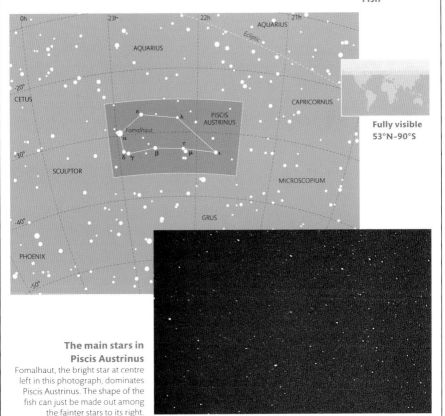

Fully visible 53°N–90°S

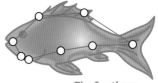

The main stars in Piscis Austrinus
Fomalhaut, the bright star at centre left in this photograph, dominates Piscis Austrinus. The shape of the fish can just be made out among the fainter stars to its right.

FEATURES OF INTEREST

Alpha (α) Piscis Austrini (Fomalhaut) ◉
A blue-white star that lies 25 light years away. At magnitude 1.2, it is not only the brightest star in the constellation but also ranks among the 20 brightest stars in the sky. The name Fomalhaut is derived from an Arabic term that means "fish's mouth", which describes the place the star occupies in the celestial pattern.

Beta (β) Piscis Austrini ✹ A wide double star, divisible with a small telescope. Its two components are of magnitudes 4.3 and 7.1.

Gamma (γ) Piscis Austrini ✹ A double star. With a small telescope it is difficult to separate the stars, as the 5th-magnitude primary greatly outshines its close 8th-magnitude neighbour.

Width:	Depth:	Area: 245 square degrees	Size ranking: 60th

Abbreviation: Pup	Genitive: Puppis	Highest in sky at 10 pm: January to February

Puppis

This rich constellation lies in the Milky Way next to Canis Major. It represents the stern of the ship of the Argonauts. Ancient Greek astronomers represented the whole ship as one constellation, Argo Navis. This was later divided into three parts, with Puppis being the largest section. Because it is only part of a once-larger constellation, Puppis has no stars labelled Alpha (α), Beta (β), Gamma (γ), Delta (δ), or Epsilon (ε).

The Stern

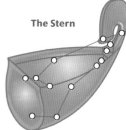

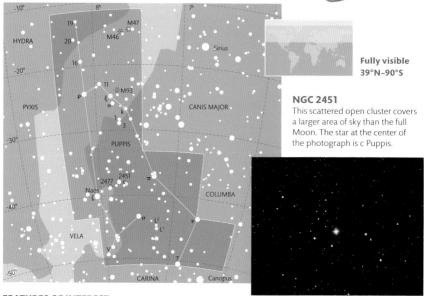

**Fully visible
39°N–90°S**

NGC 2451

This scattered open cluster covers a larger area of sky than the full Moon. The star at the center of the photograph is c Puppis.

FEATURES OF INTEREST

Zeta (ζ) Puppis (Naos) 👁 The brightest star in the constellation, at magnitude 2.2. It is a hot and luminous blue-white supergiant, lying just over 1,000 light years away. Its name is derived from a Greek word meaning "ship".

k Puppis ✶ A double star. A small telescope reveals two similar blue-white stars of almost equal brightness, at magnitudes 4.4 and 4.5.

L Puppis 👁 🔭 A wide pair of unrelated stars, each visible to the naked eye. L^1 Puppis is a blue-white star of magnitude 4.9, while L^2 is a red giant that varies between about 3rd and 6th magnitudes every 5 months or so.

V Puppis 👁 🔭 An eclipsing binary that varies between magnitudes 4.4 and 4.9 every 1 day 11 hours.

M46 🔭 ✶ An open cluster, just visible to the naked eye and appearing the same size as the full Moon. A small telescope resolves its stars. It lies over 5,000 light years away.

M47 🔭 A scattered open cluster, visible to the naked eye and slightly larger than M46 but much closer, about 1,500 light years away. Its brightest stars are of 6th magnitude.

NGC 2451 👁 🔭 An open cluster, visible to the naked eye and through binoculars. Its brightest member is an orange giant, c Puppis, of magnitude 3.6.

NGC 2477 🔭 ✶ An open cluster, 4,700 light years away. It is so rich that it resembles a globular cluster when seen through binoculars. A telescope will resolve individual stars.

Width:	Depth:	Area: 673 square degrees	Size ranking: 20th

| Abbreviation: Pyx | Genitive: Pyxidis | Highest in sky at 10 pm: February to March |

Pyxis

**Fully visible
52°N–90°S**

This southern constellation lies on the edge of the Milky Way between Hydra and Puppis. Invented by Nicolas Louis de Lacaille in the 18th century, it represents a magnetic compass of the type used by mariners.

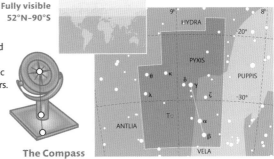

FEATURES OF INTEREST

T Pyxidis 🔭 A nova that has flared to 6th or 7th magnitude on six known occasions since 1890, the last in 2011.

The Compass

| Width: 🖐 | Depth: 🖐🖐 | Area: 221 square degrees | Size ranking: 65th |

| Abbreviation: Ret | Genitive: Reticuli | Highest in sky at 10 pm: December |

Reticulum

**Fully visible
23°N–90°S**

This small southern constellation near the Large Magellanic Cloud dates from the 18th century. It depicts a device called a grid or reticle, which was used in telescope eyepieces for recording star positions. The brightest star in the constellation is Alpha (α) Reticuli, magnitude 3.4.

The Net

FEATURES OF INTEREST

Zeta (ζ) Reticuli 👁🔭 A double star, divisible with the naked eye. The two 5th-magnitude yellow stars, similar to the Sun, both lie 39 light years away.

| Width: 🖐 | Depth: 🖐 | Area: 114 square degrees | Size ranking: 82nd |

| Abbreviation: Sge | Genitive: Sagittae | Highest in sky at 10 pm: August |

Sagitta

The Arrow   **Fully visible
90°N–69°S**

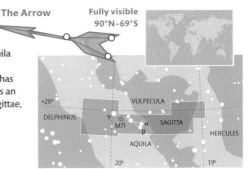

Although it lies in the Milky Way, with Vulpecula and Cygnus to the north and Aquila to the south, Sagitta contains little that is of interest. The third-smallest constellation, it has been visualized since ancient Greek times as an arrow, with its brightest star, Gamma (γ) Sagittae, magnitude 3.5, marking the arrowhead.

FEATURES OF INTEREST

M71 🔭✦ A small globular cluster, visible with binoculars or a small telescope.

| Width: 🖐🖐 | Depth: 🖐 | Area: 80 square degrees | Size ranking: 86th |

| Abbreviation: Sgr | Genitive: Sagittarii | Highest in sky at 10 pm: July to August |

Sagittarius

This constellation of the zodiac lies between Scorpius
and Capricornus. It depicts Crotus, the son of the Greek
god Pan and the inventor of archery, aiming his bow at
a scorpion, represented by the neighbouring constellation
of Scorpius. The centre of our Galaxy lies in the same direction
as Sagittarius, and binoculars show that the star fields in this part
of the Milky Way are particularly dense. The Sun passes through
Sagittarius from 18 December to 19 January, a period that includes
the December solstice. The constellation's brightest star is Epsilon
(ε) Sagittarii, magnitude 1.8.

The Archer

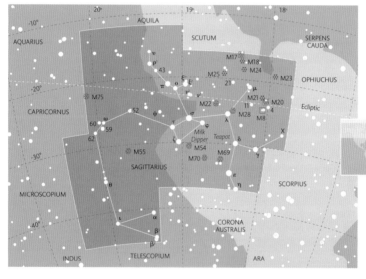

**Fully visible
44°N–90°S**

FEATURES OF INTEREST

The Teapot  A group of eight stars: Gamma (γ),
Epsilon (ε), Delta (δ), Lambda (λ), Phi (φ), Zeta (ζ),
Sigma (σ), and Tau (τ) Sagittarii. Together, they form
the outline of a teapot, with a pointed lid and a large
spout. The stars Lambda (λ), Phi (φ), Zeta (ζ), Sigma
(σ), and Tau (τ) Sagittarii also form a shape known
as the Milk Dipper, so named because it lies in a
rich area of the Milky Way.

Beta (β) Sagittarii A multiple star, although
all the components lie at different distances and
are unrelated. The naked eye shows it as a wide
double with components of magnitudes 4.0
and 4.3. The brighter star, Beta-1 (β^1), has a
7th-magnitude companion that can be seen
through a small telescope.

The main stars in Sagittarius
The Teapot is at the centre in this picture. Also visible
are M7 (an open cluster in Scorpius) at centre right,
and the arc of Corona Australis below the Teapot.

| Width: | Depth: | Area: 867 square degrees | Size ranking: 14th |

M8 (The Lagoon Nebula)
This bright nebula lies in the Milky Way in Sagittarius. The curving lane of dust after which the nebula is named is at the centre of the photograph. Immediately to the left of the dust lane is a sprinkling of stars that belong to the cluster NGC 6530, while the star enveloped in nebulosity to its right is 9 Sagittarii.

M8 (The Lagoon Nebula) ◈ ♏ ✦ A bright nebula, visible to the naked eye and a good target for binoculars. Elongated in shape, it is almost three times the apparent width of the full Moon. A dark lane of dust bisects it: one half contains the cluster NGC 6530, with stars of 7th magnitude and fainter, while the most prominent object in the other half is the 6th-magnitude star 9 Sagittarii. The nebula, and the stars within it, are 4,300 light years away.

M17 (The Omega Nebula) ♏ ✦ A cloud of glowing gas of similar apparent size to the full Moon. It can be seen through binoculars, although a telescope is needed to see its true shape. A small telescope shows a star cluster, NGC 6618, within the cloud. The nebula and cluster both lie about 5,500 light years away.

M20 (The Trifid Nebula) ✦ A nebula with a faint double star at its centre, both visible with a small telescope. Larger apertures, and photographs, show dust lanes trisecting the nebula.

M22 ♏ ✦ The third-brightest globular cluster in the sky, 10,000 light years away. It is just visible to the naked eye under good conditions and easy to find with binoculars, which show it as a rounded patch about two-thirds the apparent size of the full Moon. A telescope of moderate aperture resolves its brightest stars.

M23 ♏ ✦ A rich open cluster, elongated in shape and almost the same apparent width as the full Moon. It can be seen with binoculars, while a small telescope will resolve its stars. It lies 2,100 light years away.

M24 ◈ ♏ ✦ A large and bright field of stars in the Milky Way, visible to the naked eye and an excellent sight through binoculars. A small telescope reveals a small open cluster, NGC 6603, within it.

M25 ♏ ✦ An open cluster, just visible to the naked eye and a good object for observation with binoculars or a small telescope. Its stars, of 7th magnitude and fainter, are scattered over an area of sky the same apparent size as the full Moon. It lies 2,000 light years away.

M17 (The Omega Nebula)
When viewed with a telescope, this nebula resembles the capital letter Omega (Ω) in the Greek alphabet.

M20 (The Trifid Nebula)
A telescope reveals the three main lanes of dust that cross this nebula and give rise to its popular name.

Abbreviation: Sco	Genitive: Scorpii	Highest in sky at 10 pm: June to July

Scorpius

A constellation of the zodiac, Scorpius lies between Libra and
Sagittarius. It depicts the scorpion that, in Greek mythology, killed
Orion with its sting – fittingly, Orion sets as Scorpius rises. The
constellation lies in a rich region of the Milky Way, in the same
direction as the centre of our Galaxy. The Sun passes through
it briefly, from 23 to 29 November. The old version of its name,
Scorpio, is used only in astrology.

The Scorpion

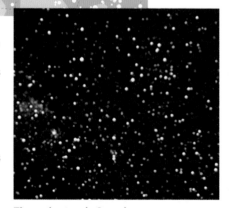

**Fully visible
44°N–90°S**

FEATURES OF INTEREST

Alpha (α) Scorpii (Antares) ◉ ♫ A red supergiant of
variable brightness, ranging between magnitudes 0.9
and 1.2 in a cycle lasting approximately 4–5 years. It has
a close blue-white companion of 5th magnitude that
orbits it every 2,500 years or so and can be seen with
a telescope of moderate aperture. The name Antares
is usually translated as "rival of Mars", referring to its
reddish colour, although it can also mean "like Mars".

Beta (β) Scorpii ☇ A double star, which has components
of magnitudes 2.6 and 4.9 that are easy to separate with
a small telescope. The two stars lie just over 400 light
years away, but are not thought to orbit each other.

Zeta (ζ) Scorpii ◉ ♫ A wide pair of unrelated stars,
divisible with the naked eye by those with good
eyesight. Zeta-1 (ζ¹) is a blue-white supergiant of

The main stars in Scorpius
The stars representing the scorpion's head are at the
top right here. Orange Antares marks the heart, and
the stars in the body and tail curve down to the left.

Width:	Depth:	Area: 497 square degrees	Size ranking: 33rd

Nebulosity around Antares
A faint area of nebulosity, visible only on long-exposure photographs, extends northwards from Antares (below centre in this picture) to Rho (ρ) Ophiuchi (top centre).

magnitude 4.7, the brightest star in the cluster NGC 6231 (see below). Zeta-2 (ζ²), a red giant of magnitude 3.6, is much closer at 130 light years away.

Mu (μ) Scorpii ◉ � A double star that can be divided with the naked eye. The brighter component is an eclipsing binary that varies between magnitudes 2.9 and 3.2 every 1 day 10 hours. Its companion is of magnitude 3.5.

Nu (ν) Scorpii ☄ A multiple star. A small telescope, or even good binoculars, will reveal an optical double, with components of magnitudes 4.3 and 6.6. The fainter star of the pair has an 8th-magnitude companion that can be seen through a telescope

with an aperture of 75mm (3in). Larger apertures – 100mm (4in) or more – show that the brighter star has an even closer companion, of 5th magnitude. Hence Nu Scorpii is an apparent quadruple.

Xi (ξ) Scorpii ☄ A multiple star. Like Nu (ν) Scorpii, this is a quadruple. A small telescope reveals a pair of 5th- and 7th-magnitude stars; in the same field of view, a wider pair of 7th- and 8th-magnitude stars can also be seen.

Omega (ω) Scorpii ◉ ♩ A naked-eye double star, with components of magnitudes 4.0 and 4.3, 479 and 290 light years away.

M4 ♩☄ One of the closest globular clusters to us, about 6,500 light years away. It can be seen with binoculars or a small telescope, but a dark sky is needed as its light is spread over a large area, two-thirds the apparent size of the Moon.

M6 ◉ ♩☄ An open cluster, 1,500 light years away. It is visible to the naked eye, and its individual stars can be seen with binoculars. Its brightest star is BM Scorpii, an orange giant that varies between 5th and 7th magnitudes.

M7 ◉ ♩ A large, glorious open cluster, visible with the naked eye and binoculars and more than twice the apparent width of the full Moon. Its brightest stars, of 6th magnitude, are seen against a bright Milky Way background. The cluster is 980 light years away.

NGC 6231 ♩☄ A prominent open cluster, 5,500 light years away. Its individual stars are easy to see with binoculars or a small telescope. The 5th-magnitude star Zeta-1 (ζ¹) Scorpii (see above) is its brightest member.

M4
This globular cluster, between Antares and Sigma (σ) Scorpii, is large but faint and hence difficult to see.

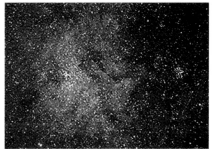

M6 and M7
M6 is at centre right in this picture, with M7 to its left in front of the bright band of the Milky Way.

| Abbreviation: Scl | Genitive: Sculptoris | Highest in sky at 10 pm: October to November |

Sculptor

This faint constellation is to the south of Aquarius and Cetus. It was invented in the 18th century by the French astronomer Nicolas Louis de Lacaille, who visualized it as a sculptor's studio, although its name has since been changed. Its brightest star, Alpha (α) Sculptoris, is of only magnitude 4.3. Sculptor contains the south pole of our Galaxy – the point 90 degrees south of the plane of the Milky Way. In this region, our view into the Universe is uninterrupted by intervening gas and dust in our Galaxy, so many faint and distant galaxies are visible. The north galactic pole lies in Coma Berenices.

Sculptor

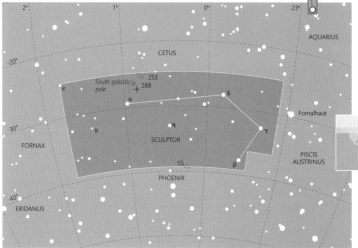

FEATURES OF INTEREST

Epsilon (ε) Sculptoris ✴ A double star that can be divided with a small telescope. The 5th- and 9th-magnitude components are a true binary pair that orbit each other every 2,000 years or so. They are 91 light years away.

R Sculptoris ♠ A variable red giant that ranges between 6th and 8th magnitudes in a cycle lasting about a year.

NGC 55 ✴ A spiral galaxy, tilted nearly edge-on to us, so that it appears cigar-shaped when viewed with a small telescope.

NGC 253 ♠✴ A spiral galaxy. Like NGC 55, it appears to us edge-on but it is brighter and hence visible with binoculars. A telescope of moderate to large aperture shows traces of dark dust clouds silhouetted against its stars.

NGC 253 and NGC 288
These two objects are in the northern part of Sculptor. NGC 288 (on the left in this picture) is a globular cluster that can be seen through a telescope. The spiral galaxy NGC 253 is on the right.

| Width: | Depth: | Area: 475 square degrees | Size ranking: 36th |

Scutum

Scutum is a small constellation that lies just south of the celestial equator in a rich area of the Milky Way between Aquila and Sagittarius. It was introduced in the late 17th century by the Polish astronomer Johannes Hevelius, who originally termed it Sobieski's Shield in honour of his patron, King John Sobieski. Its brightest star is Alpha (α) Scuti, magnitude 3.8.

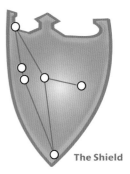

The Shield

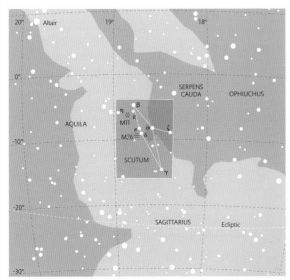

**Fully visible
74°N–90°S**

M11 (The Wild Duck Cluster)

The open cluster M11 lies in a bright section of the Milky Way near the border between Aquila and Scutum. On this photograph, taken through a telescope, it is just possible to see the V-shape, with the apex at the top right of the cluster.

FEATURES OF INTEREST

Delta (δ) Scuti 👁 ♏ A pulsating giant, the prototype of a class of variable stars that show very small fluctuations in brightness (a few tenths of a magnitude or less) over short periods of time (no more than a few hours). Delta Scuti itself varies between magnitudes 4.6 and 4.8 over a period of less than 5 hours. It lies 200 light years away.

R Scuti ♏ A pulsating orange supergiant, which varies between magnitudes 4.2 and 8.6 in a 20-week cycle.

M11 (The Wild Duck Cluster) ♏⚹ A rich open cluster, consisting of hundreds of stars, visible in binoculars as a fuzzy ball nearly half the apparent size of the full Moon. In a small telescope it appears V-shaped, like a flight of ducks, hence its popular name. Near the apex of the V is the cluster's brightest star, of 8th magnitude. M11 is about 6,000 light years

away and lies on the northern edge of a particularly bright section of the Milky Way known as the Scutum Star Cloud.

Width: 👆	Depth: 👆	Area: 475 square degrees	Size ranking: 84th

Abbreviation: Ser	Genitive: Serpentis	Highest in sky at 10 pm: June to August

Serpens

Uniquely, Serpens consists of two separate areas that, taken together, are regarded as one constellation. It depicts a serpent coiled around Ophiuchus, who holds the head (Serpens Caput) in his left hand and the tail (Serpens Cauda) in his right. The constellation's brightest star, Alpha (α) Serpentis, magnitude 2.6, is also known as Unukalhai, a name derived from an Arabic term that means "the serpent's neck".

The Serpent

**Fully visible
74°N–64°S**

FEATURES OF INTEREST

Delta (δ) Serpentis ✶ A binary star. The 4th-magnitude primary and its 5th-magnitude partner can be divided with a small telescope and high magnification. Their orbital period is estimated to be around 1,200 years.

Theta (θ) Serpentis ✶ A wide double star. The components, of magnitudes 4.6 and 5.0, can be separated through a small telescope.

Nu (ν) Serpentis ♙✶ A wide double, with 4th- and 9th-magnitude components that are divisible with binoculars or a small telescope.

Tau-1 (τ¹) Serpentis ♙ A 5th-magnitude star, the brightest member of a triangular field of stars that fans out towards Iota (ι) and Kappa (κ) Serpentis.

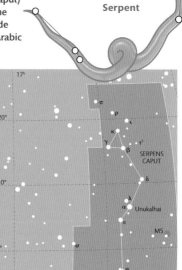

M5
This globular cluster in the southern part of Serpens Caput is visible through binoculars. It is one of the finest objects of its type in northern skies.

Width:	Depth:	Area: 637 square degrees	Size ranking: 23rd

M5 ♏ ↗ A globular cluster, about 25,000 light years away. It can be seen through binoculars, appearing about half the size of the full Moon. A telescope of moderate aperture resolves its brightest stars; it also reveals a condensed centre and chains of stars in the outer regions.

M16 ♏ ↗ An open cluster, visible through binoculars or a small telescope. It appears hazy, since the stars are embedded in the Eagle Nebula, which shows up well only with larger apertures and on photographs. Cluster and nebula are around 5,700 light years away.

The Eagle Nebula
This nebula can be seen through a telescope of large aperture or on a long-exposure photograph.

Dark columns in the Eagle Nebula
This close-up, taken by the Hubble Space Telescope, shows dark fingers of dust against areas of brighter gas.

Abbreviation: Sex	Genitive: Sextantis	Highest in sky at 10 pm: March to April

Sextans

This faint constellation is on the celestial equator, south of Leo. It was invented in the late 17th century by the Polish astronomer Johannes Hevelius and represents the sextant with which he measured star positions. Its brightest star is Alpha (α) Sextantis, magnitude 4.5.

The Sextant

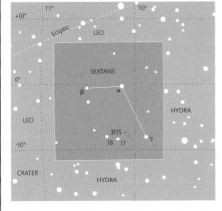

Fully visible 78°N–83°S

FEATURES OF INTEREST

17 and 18 Sextantis ◈ ♏ A wide pair of unrelated stars, detectable with the naked eye by those with good eyesight and easy to divide with binoculars. They are of magnitudes 5.9 and 5.6, and lie 600 and 520 light years away.

NGC 3115 (The Spindle Galaxy) ↗
An elliptical galaxy, visible through a telescope of small to moderate aperture. Its elongated shape gives rise to its popular name.

Width: 🖐	Depth: 🖐	Area: 314 square degrees	Size ranking: 47th

Taurus

This imposing constellation of the zodiac lies between Aries and Gemini. It represents the bull into which the Greek god Zeus transformed himself to abduct Princess Europa of Phoenicia. Zeus then swam to Crete with the princess on his back. The constellation represents the front half of the bull's body – the part visible above the Mediterranean waves. It contains two major star clusters, the Pleiades and Hyades. In mythology, the Pleiades were the seven daughters of Atlas and Pleione, and the cluster is also known as the Seven Sisters; the Hyades were the daughters of Atlas and Aethra. In the sky, the Hyades cluster marks the bull's face, while the red giant star Aldebaran forms the creature's bloodshot eye. The tips of the bull's horns are marked by Beta (β) and Zeta (ζ) Tauri, magnitudes 1.7 and 3.0. The Sun passes through Taurus from 14 May to 21 June.

The Bull

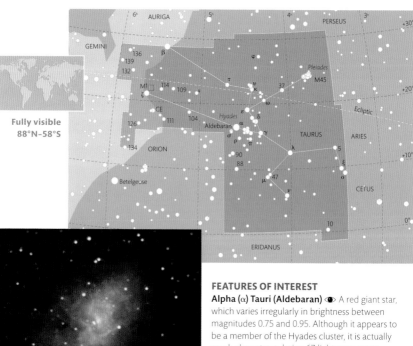

Fully visible 88°N–58°S

M1 (The Crab Nebula)
This remarkable object is the remains of a massive star that exploded as a supernova.

FEATURES OF INTEREST

Alpha (α) Tauri (Aldebaran) ◉ A red giant star, which varies irregularly in brightness between magnitudes 0.75 and 0.95. Although it appears to be a member of the Hyades cluster, it is actually much closer to us, being 67 light years away.

Theta (θ) Tauri ◉ ♏ A wide double star in the Hyades cluster. Observers with good eyesight can divide the two stars with the naked eye. Theta-1 (θ^1) is a yellow giant, magnitude 3.8; Theta-2 (θ^2) is a white giant of magnitude 3.4, the brightest member of the Hyades.

| Width: | Depth: | Area: 797 square degrees | Size ranking: 17th |

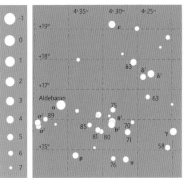

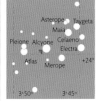

The Pleiades

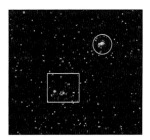

The Hyades and Pleiades
These two outstanding open clusters lie close together in the sky.

The Hyades

Magnitude key

Lambda (λ) Tauri 👁 An eclipsing binary star of the same type as Algol. It ranges between magnitudes 3.4 and 3.9 in a cycle lasting just under 4 days.

M1 (The Crab Nebula) 🌠🔭 The remains of a supernova that was seen from Earth in AD 1054. Under excellent conditions it can be found with binoculars or a small telescope, but a moderate aperture is needed to see it well. It is elliptical in shape, appearing midway in size between the disc of a planet and the full Moon. It lies about 6,500 light years away.

M45 (The Pleiades) 👁🌠🔭 A large, bright open star cluster, easy to see with the naked eye and a superb sight through binoculars, appearing almost four times wider than the full Moon. Its brightest star is Eta (η) Tauri (Alcyone), a blue-white giant of magnitude 2.9. Those with normal eyesight can see about six stars, but dozens are visible with binoculars or a small telescope. The cluster is about 440 light years away. Long-exposure photographs show nebulosity around the stars, but a large telescope is needed to see this directly.

The Hyades 👁🌠🔭 A large, loose, V-shaped star cluster, easily visible to the naked eye. It is best viewed with binoculars because of its considerable size, being scattered across the apparent width of 10 full Moons. The cluster lies about 150 light years away.

Abbreviation: Tel	Genitive: Telescopii	Highest in sky at 10 pm: July to August

Telescopium

Fully visible 33°N–90°S

This unremarkable constellation south of Sagittarius was introduced by the French astronomer Nicolas Louis de Lacaille. In its original form, it represented a large telescope supported by a winch, but it is a poor tribute to such an important instrument. Its brightest star is Alpha (α) Telescopii, magnitude 3.5.

The Telescope

FEATURES OF INTEREST

Delta (δ) Telescopii 👁🌠 An optical double star, divisible without optical aid by those with good eyesight and easy to see with binoculars. Delta-1 (δ¹), magnitude 4.9, is about 700 light years away; Delta-2 (δ²) is of magnitude 5.1 and lies just under 1,000 light years away.

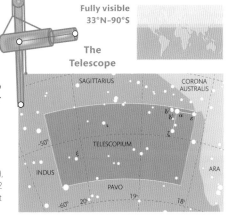

Width:	Depth:	Area: 252 square degrees	Size ranking: 57th

Abbreviation: Tri	Genitive: Trianguli	Highest in sky at 10 pm: November to December

Triangulum

This small constellation can be found between Andromeda and Aries. It was known to the ancient Greeks, many of whom visualized it as the Nile delta, while for others it represented the island of Sicily. Its brightest star is Beta (β) Trianguli, magnitude 3.0.

The Triangle

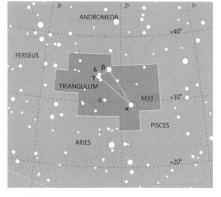

**Fully visible
90°N–52°S**

FEATURES OF INTEREST
6 Trianguli ✶ A binary star. A small telescope is needed to divide the 5th-magnitude primary from its close 7th-magnitude companion.

M33 ♠✶ A spiral galaxy, nearly 3 million light years away and the third-largest member of the Local Group of galaxies. It covers about the same area of sky as the full Moon. Because its light is so spread out, a clear, dark sky is needed to see it. Under good conditions, this galaxy can be detected through binoculars or a small telescope.

Width:	Depth:	Area: 132 square degrees	Size ranking: 78th

Abbreviation: TrA	Genitive: Trianguli Australis	Highest in sky at 10 pm: June to July

Triangulum Australe

Lying in the Milky Way near Alpha (α) and Beta (β) Centauri is the southern counterpart of the northern triangle, Triangulum. Although smaller than Triangulum, it is more prominent because its main stars are brighter. It is the smallest of the 12 constellations introduced at the end of the 16th century by the Dutch navigators Pieter Dirkszoon Keyser and Frederick de Houtman. Its brightest star is Alpha (α) Trianguli Australis, magnitude 1.9.

**The
Southern
Triangle**

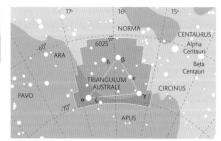

Triangulum Australe
The three corners of the triangle stand out among the fainter stars in the surrounding Milky Way.

**Fully visible
19°N–90°S**

FEATURES OF INTEREST
NGC 6025 ♠✶ An open cluster in the Milky Way, about one-third the apparent width of the full Moon and easy to see with binoculars or a small telescope, its brightest star being of 7th magnitude. It lies 2,500 light years away.

Width:	Depth:	Area: 110 square degrees	Size ranking: 83rd

| Abbreviation: Tuc | Genitive: Tucanae | Highest in sky at 10 pm: September to November |

Tucana

This southern constellation, representing a toucan, the large-beaked bird of South and Central America, lies close to the bright star Achernar and south of two other celestial birds, Grus and Phoenix. It was invented in the late 16th century by the Dutch navigators Pieter Dirkszoon Keyser and Frederick de Houtman. Its brightest star is Alpha (α) Tucanae, magnitude 2.8, but its most notable naked-eye object is the Small Magellanic Cloud.

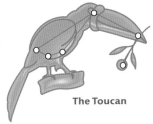

The Toucan

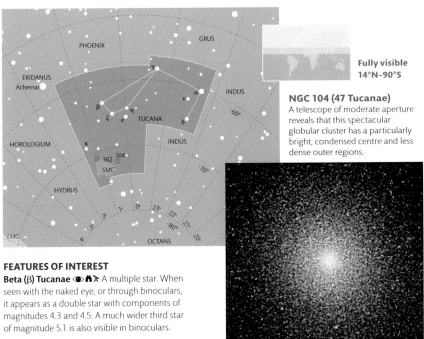

**Fully visible
14°N–90°S**

NGC 104 (47 Tucanae)
A telescope of moderate aperture reveals that this spectacular globular cluster has a particularly bright, condensed centre and less dense outer regions.

FEATURES OF INTEREST

Beta (β) Tucanae ◉▶📷☀ A multiple star. When seen with the naked eye, or through binoculars, it appears as a double star with components of magnitudes 4.3 and 4.5. A much wider third star of magnitude 5.1 is also visible in binoculars.

Kappa (κ) Tucanae ☀ A double star, with components of 5th and 8th magnitudes that can be separated through a small telescope.

NGC 104 (47 Tucanae) ◉▶📷☀ A globular cluster, rated the second-best of its type in the whole sky. To the naked eye, it appears similar to a hazy star of 4th magnitude. Binoculars or a small telescope show it covering the same area of sky as the full Moon, although an aperture of 100mm (4in) is needed to resolve its stars. The cluster lies 15,000 light years away.

NGC 362 📷☀ A globular cluster, visible with binoculars or a small telescope. It appears to

lie near the Small Magellanic Cloud (see below) but is, in fact, only 29,000 light years away and hence is in our own Galaxy.

The Small Magellanic Cloud (SMC) ◉▶📷☀ The smaller of the Milky Way's two satellite galaxies. It appears to the naked eye as an elongated patch of light seven times wider than the full Moon, while binoculars or a small telescope resolve individual clusters and nebulae within it. This galaxy is much smaller than our own, having only a few per cent our Galaxy's mass and less than 10 per cent of its diameter. It is about 190,000 light years away.

| Width: 👀 | Depth: ✋ | Area: 295 square degrees | Size ranking: 48th |

| Abbreviation: UMa | Genitive: Ursae Majoris | Highest in sky at 10 pm: February to May |

Ursa Major

Ursa Major is among the most famous constellations. It is the third-largest in the sky, occupying a much wider area than that covered by the group of seven stars that form the asterism of the Plough (or Big Dipper). Ursa Major represents Callisto, a mortal in Greek myth and a hunting partner of Artemis, who was seduced by Zeus. In different versions of the story, she was turned into a bear either by Zeus's jealous wife, Hera, or by the angry Artemis. Its brightest stars are Alpha (α) and Epsilon (ε) Ursae Majoris, both magnitude 1.8. A line drawn from Beta (β) through Alpha (α) Ursae Majoris points towards Polaris, the north Pole Star, in adjacent Ursa Minor.

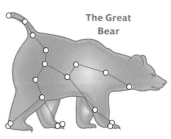

The Great Bear

**Fully visible
90°N–16°S**

FEATURES OF INTEREST

The Plough (The Big Dipper) ◄●► One of the best-known patterns in the sky, consisting of the naked-eye stars Alpha (α), Beta (β), Gamma (γ), Delta (δ), Epsilon (ε), Zeta (ζ), and Eta (η) Ursae Majoris. The outline that they form is visualized in various cultures as that of a plough, a ladle, a saucepan, and even a wagon. All the stars in the

Plough, with the exception of the outermost two (Alpha and Eta), are travelling in the same direction through space and form what is known as a moving cluster.

Zeta (ζ) Ursae Majoris (Mizar) ◄●► A multiple star. Observers with good eyesight or using binoculars will see that this star, of magnitude 2.2, has a partner

| Width: | Depth: | Area: 1,280 square degrees | Size ranking: 3rd |

Ursa Major (Inset: Mizar and Alcor)
The main photograph shows the whole of the Great Bear, with the Plough (or Big Dipper) forming the animal's rump and tail. Mizar, the second star in the handle of the Plough (at the right of the inset picture), forms an optical double with Alcor (left of inset) and a true binary with its closer companion.

of magnitude 4.0, known as Alcor or 80 Ursae Majoris. Mizar and Alcor are both just over 80 light years away but are not a true binary pair. However, a small telescope reveals that Mizar has a closer 4th-magnitude companion. This pair forms a true binary with a very long orbital period.

Xi (ξ) Ursae Majoris ✷ A binary star, with close components of magnitudes 4.3 and 4.8 that can be separated through a telescope of 75mm (3in) aperture. The yellowish stars, similar to the Sun, are 28 light years away and have an orbital period of 60 years.

M81 and M82 ♙✷ Two contrasting galaxies, about 12 million light years away. M81 is a beautiful spiral, visible with binoculars or a small telescope. One Moon diameter to the north is the smaller and fainter M82, a galaxy of peculiar appearance, which is now thought to be a spiral seen edge-on that is passing through a dust cloud.

M101 ♙✷ A spiral galaxy, which appears face-on to us. It covers almost as much sky as the full Moon but it is quite faint, and good conditions are required if it is to be seen with binoculars or a small telescope.

M81
This beautiful spiral galaxy is one of the easiest examples of its type to see with a small telescope.

M101
The arms of this spiral galaxy can be seen with a large telescope or on a long-exposure photograph.

| Abbreviation: UMi | Genitive: Ursae Minoris | Highest in sky at 10 pm: May to June |

Ursa Minor

Ursa Minor contains the north celestial pole. By chance, there is a moderately bright star, known as Polaris (or the north Pole Star), about 1 degree from the pole. Navigators have long recognized that when looking at Polaris one is facing almost due north. Ursa Minor's main stars form a shape known as the Little Dipper, while the stars Beta (β) and Gamma (γ) Ursae Minoris (Kochab and Pherkad), in the dipper's bowl, are known as the Guardians of the Pole. Ursa Minor is named after a nymph who nursed the infant Zeus, although it is not clear why she is depicted as a bear.

The Lesser Bear

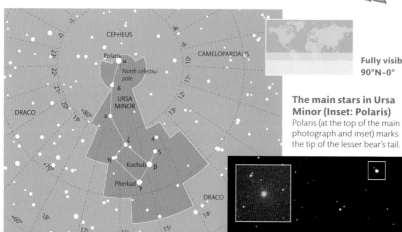

Fully visible
90°N–0°

The main stars in Ursa Minor (Inset: Polaris)
Polaris (at the top of the main photograph and inset) marks the tip of the lesser bear's tail.

FEATURES OF INTEREST

Alpha (α) Ursae Minoris (Polaris) ◉ ⋇
The north Pole Star and the brightest star in the constellation, a yellow-white supergiant of magnitude 2.0. It is in fact a Cepheid variable, although its fluctuations in magnitude are so slight that they cannot be detected with the naked eye. A small telescope shows an unrelated 8th-magnitude companion.

Gamma (γ) Ursae Minoris (Pherkad) ◉ ♙
An optical double star, with components that are divisible with the naked eye or binoculars. Gamma itself is a blue-white giant star of magnitude 3.0, 490 light years away. Its partner, 11 Ursae Minoris, is an orange giant, of magnitude 5.0, 410 light years away.

Eta (η) Ursae Minoris ◉ ♙ An optical double star. It is a white star of magnitude 5.0, lying 96 light years away. The naked eye and binoculars show a wide companion, 19 Ursae Minoris, of magnitude 5.5, which is over six times further away than the primary.

| Width: | Depth: | Area: 256 square degrees | Size ranking: 56th |

Abbreviation: Vel	Genitive: Velorum	Highest in sky at 10 pm: February to April

Vela

This is one of the three parts into which the ancient Greek constellation of Argo Navis (the ship of the argonauts) was divided by Nicolas Louis de Lacaille. It represents the sails of the ship. Vela lies in the Milky Way with Carina and Puppis, the other two portions of the ship, on one side, and Centaurus on the other. Since it is only a part of the once-larger constellation, Vela contains no stars labelled Alpha (α) or Beta (β).

The Sails

Fully visible 32°N–90°S

NGC 3132 (The Eight-Burst Nebula)

Long-exposure photographs such as this one reveal the loops of gas, intertwined like figures of eight, after which this nebula is named.

FEATURES OF INTEREST

Gamma (γ) Velorum ♠ ⚹ A multiple star. Through a small telescope or even good binoculars, it appears as a wide double star. The primary, Gamma-2 (γ²), is the brightest star in the constellation, at magnitude 1.8. It is a spectroscopic binary, one component of which is a Wolf-Rayet star, a rare type of ultra-hot star that has lost its outer hydrogen layers. The primary's visible partner, Gamma-1 (γ¹), is an ordinary blue-white star of magnitude 4.2. There are also two wider companions of 7th and 9th magnitudes, which are visible through a small telescope.

NGC 2547 ♠ ⚹ An open cluster, visible with binoculars or a small telescope. It is more than half the apparent size of the full Moon and lies 1,200 light years away.

NGC 3132 (The Eight-Burst Nebula) ⚹ A planetary nebula. Its rounded disc is visible with a small telescope, which may also reveal the 10th-magnitude star at its centre. Photographs show complex loops within the nebula.

IC 2391 👁 ♠ ⚹ A large open cluster, visible to the naked eye and an ideal object for binocular observation, covering a greater area of sky than the full Moon. Its brightest star is Omicron (o) Velorum, magnitude 3.6. The cluster lies about 500 light years away.

Width:	Depth:	Area: 500 square degrees	Size ranking: 32nd

Abbreviation: Vir	Genitive: Virginis	Highest in sky at 10 pm: April to June

Virgo

This is the largest constellation of the zodiac and the second largest of all the constellations. It lies on the celestial equator between Leo and Libra. The constellation is usually identified as Dike, the Greek goddess of justice, or sometimes as Demeter, the corn goddess. Virgo is of particular interest because it contains the nearest large cluster of galaxies, the Virgo Cluster. The Sun passes through the constellation from 16 September to 31 October.

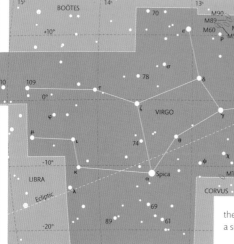

The Virgin

Fully visible 67°N–75°S

FEATURES OF INTEREST

Alpha (α) Virginis (Spica) ◕ The brightest star in Virgo and, at magnitude 1.0, among the 20 brightest in the sky. It is a blue-white star, 250 light years away. Spica is Latin for "ear of wheat", an object Virgo holds in her left hand.

Gamma (γ) Virginis ⚹ A binary star. To the naked eye, it looks like one star of magnitude 2.7. In fact, there are two yellow-white stars (both of magnitude 3.5) orbiting a common centre of mass every 169 years. This motion affects the binary's appearance significantly. In 2005, when the stars were closest together as seen from Earth, an aperture of 250mm (10in) was needed to divide them. After that they moved rapidly apart. Currently the stars are divisible with a small aperture and will remain that way for the rest of the century.

M87 ⚹ A famous giant elliptical galaxy near the centre of the Virgo Cluster (see below). It is probably the easiest member of the cluster to see through a small telescope.

M104 (The Sombrero Galaxy) ⚹ A spiral galaxy that appears almost edge-on to us and looks elongated when seen with a small telescope. Larger apertures show a dark lane of dust in the spiral arms, crossing the central nucleus of stars. The galaxy is not part of the Virgo Cluster, but lies somewhat closer to us.

The Virgo Cluster (The Virgo-Coma Cluster) ⚹ A cluster of 2,000 or more galaxies, about 50 million light years away, that extends from Virgo into Coma Berenices. Its brightest members, notably the elliptical galaxies M49, M60, M84, M86, and M87, are visible with the size of telescope used by amateur observers.

M104 (The Sombrero Galaxy) This spiral galaxy owes its name to its flattened appearance.

Width:	Depth:	Area: 1,294 square degrees	Size ranking: 2nd

Abbreviation: Vol	Genitive: Volantis	Highest in sky at 10 pm: January to March

Volans

This faint southern constellation adjoins Carina. Invented at the end of the 16th century by the Dutch navigators Pieter Dirkszoon Keyser and Frederick de Houtman, it represents the fish of tropical waters that can glide above the waves on its outstretched fins.

The Flying Fish

Fully visible
14°N–90°S

FEATURES OF INTEREST

Gamma (γ) Volantis ✶ An attractive double star. A small telescope reveals its orange and yellow components, of magnitudes 3.8 and 5.6.

Epsilon (ε) Volantis ✶ A double star with components of magnitude 4.4 and 7.3 that can be divided with a small telescope.

Width:	Depth:	Area: 141 square degrees	Size ranking: 76th

Abbreviation: Vul	Genitive: Vulpeculae	Highest in sky at 10 pm: August to September

Vulpecula

Vulpecula is in the Milky Way, to the south of Cygnus. When it was introduced in the late 17th century by the Polish astronomer Johannes Hevelius, it was named Vulpecula cum Anser (the Fox and Goose). Its brightest star is Alpha (α) Vulpeculae, magnitude 4.4.

The Fox

Fully visible
90°N–61°S

Brocchi's Cluster
This open cluster lies in the southern part of Vulpecula, within the division in the Milky Way known as the Cygnus Rift.

FEATURES OF INTEREST

M27 (The Dumbbell Nebula) ♏✶ A large planetary nebula, appearing through binoculars as a rounded patch about a quarter the size of the full Moon. Its twin-lobed shape, which explains its popular name, can be seen with a telescope of moderate to large aperture or on a photograph. It is about 1,200 light years away.

Brocchi's Cluster (Collider 399, The Coathanger) ♏
An unusual open cluster that is easy to find with binoculars. It consists of a line of six stars of 6th and 7th magnitudes, from the centre of which extends a "hook" formed by four other stars, giving it the appearance of a coathanger.

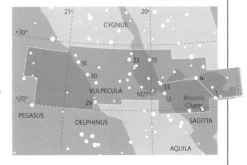

Width:	Depth:	Area: 268 square degrees	Size ranking: 55th

MONTHLY SKY GUIDES

HOW THIS SECTION WORKS

THIS SECTION INCLUDES detailed sky charts for every month of the year, with a separate chart for each hemisphere. Each month is introduced by a double-page feature, which looks at interesting celestial phenomena and uses simplified charts to assist in identifying prominent stars and constellations.

USING THE CHARTS

Both the introductory charts (see below) and the whole-sky charts (see opposite) show the night sky as it appears from latitudes 60–20°N (northern hemisphere) and 0–40°S (southern hemisphere). Use the world map (right) to identify your own latitude and associated colour code, then decide which hemisphere map is best for you. You can then use the colour-coded lines and crosses on all charts to identify the viewing horizon and zenith (point overhead) for your latitude. Compared with the whole-sky charts, the introductory charts show a simplified and less detailed view of the night sky.

Finding your colour code

Use this map to identify the colour of the latitude line nearest to your geographical location. Bear in mind that 10 degrees difference in latitude will have little effect on the stars you can actually see.

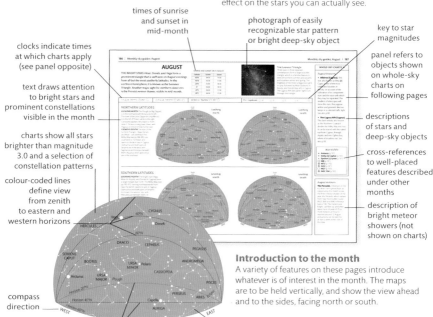

clocks indicate times at which charts apply (see panel opposite)

text draws attention to bright stars and prominent constellations visible in the month

charts show all stars brighter than magnitude 3.0 and a selection of constellation patterns

colour-coded lines define view from zenith to eastern and western horizons

compass direction

colour-coded horizon lines

times of sunrise and sunset in mid-month

photograph of easily recognizable star pattern or bright deep-sky object

key to star magnitudes

panel refers to objects shown on whole-sky charts on following pages

descriptions of stars and deep-sky objects

cross-references to well-placed features described under other months

description of bright meteor showers (not shown on charts)

Introduction to the month

A variety of features on these pages introduce whatever is of interest in the month. The maps are to be held vertically, and show the view ahead and to the sides, facing north or south.

ecliptic

How to use the whole-sky charts

To view the sky to the south, hold the book flat with the SOUTH label (at the edge of the chart) nearest your body. Your colour-coded line (around the near edge of the chart) indicates the horizon in front of you; the cross near the chart centre represents the zenith (the point directly overhead). The area of the chart beyond your zenith represents the sky to the north; to view this, turn the book around.

when looking south, hold right-hand edge of book nearest to you

STAR-MOTION DIAGRAMS

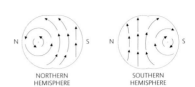

NORTHERN HEMISPHERE

SOUTHERN HEMISPHERE

Next to each whole-sky chart is a diagram showing the direction in which the stars appear to move as the night progresses. The arrows show a general motion from east to west, due to the Earth's rotation, but circumpolar stars (at one end of each chart, as shown by the diagram) rotate around one of the celestial poles without setting.

Whole-sky chart

chart shows all stars brighter than magnitude 5.0

all 88 constellation patterns are plotted on the charts

symbol for deep-sky object

colour-coded crosses indicate zeniths from various latitudes

compass direction

colour-coded horizon lines for use when facing north

key to star magnitudes

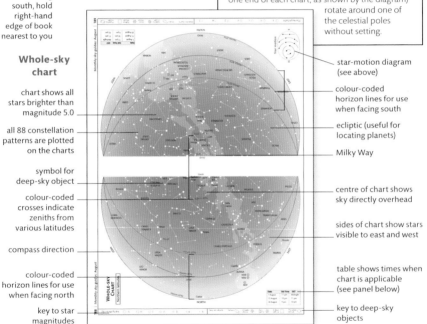

star-motion diagram (see above)

colour-coded horizon lines for use when facing south

ecliptic (useful for locating planets)

Milky Way

centre of chart shows sky directly overhead

sides of chart show stars visible to east and west

table shows times when chart is applicable (see panel below)

key to deep-sky objects

TIME

The charts for a particular month show the sky as it appears at 10pm in mid-month – that is, at 10pm standard (Std) time for a particular time zone. When daylight-saving time (DST or summer time) is in use, they show the sky as it appears one hour later. The sky will look the same at 11pm at the start of the month and at 9pm at the end of the month

as it does at 10pm in mid-month. If you wish to look at the sky at a different time, you will need to use charts shown under a different month: for every two hours of time before or after 10pm, go one month backwards or forwards. So, if you want to look at the sky at midnight on 15 January, turn to the chart for February.

JANUARY

JANUARY EVENING SKIES are the most magnificent of the year, with Orion surrounded by a host of bright stars, including Sirius, the brightest of all. From southern latitudes, Canopus, the second-brightest star, and the nearby small galaxy known as the Large Magellanic Cloud add to the richness of the scene.

SUNRISE AND SUNSET ON 15 JANUARY

Latitude	Sunrise	Sunset
60°N	08.50	15.30
40°N	07.20	17.00
20°N	06.40	17.40
0°	06.10	18.10
20°S	05.30	18.50
40°S	04.50	19.30

1 January: Std time ◷ DST ◷	15 January: Std time ◷ DST ◷	30 January: Std time ◷ DST ◷

NORTHERN LATITUDES

LOOKING NORTH The saucepan shape of the Plough (or Big Dipper) stands on its handle to the right of Polaris, the Pole Star. The W-shape of Cassiopeia is to Polaris's left, with Cepheus beneath it. Capella is almost overhead. Leo rises in the northeast, while the Great Square of Pegasus sets in the northwest.

LOOKING SOUTH Magnificent Orion commands centre stage, to the right of his dogs, marked by Sirius (in Canis Major) and Procyon (in Canis Minor). Aldebaran glints above and to the right of Orion, with the Pleiades cluster (see p.147) higher and further to the right. Nearly overhead are Castor and Pollux, along with Capella.

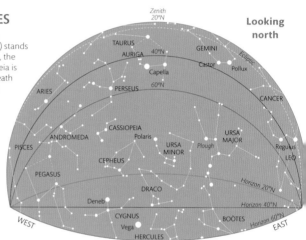

SOUTHERN LATITUDES

LOOKING NORTH Orion stands high, with Capella and the other stars of Auriga closer to the northern horizon. Brilliant Sirius and Procyon are visible to Orion's right. Aldebaran and the stars of Taurus are to the lower left of Orion, with Castor and Pollux to Orion's lower right, in the northeast. Perseus sets in the northwest.

LOOKING SOUTH Canopus is high in the south, with Sirius above it, almost overhead. Lower down, to the southwest, Achernar marks the end of the celestial river, Eridanus. Just below Canopus is the Large Magellanic Cloud (LMC); the Small Magellanic Cloud (SMC) is to its lower right (see p.149).

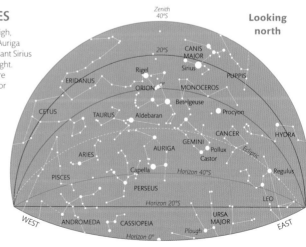

Orion's Belt
The three stars at the centre of this picture form a distinctive line that marks the belt of Orion, the hunter (pp.114–15). They are (from right to left): Delta (δ) Orionis (also known as Mintaka), Epsilon (ε) Orionis (Alnilam), and Zeta (ζ) Orionis (Alnitak).

Star magnitudes ⦿ -1 ◯ 0 ◯ 1 ◦ 2 ∘ 3

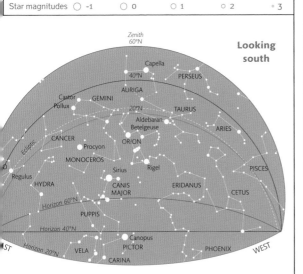

Looking south

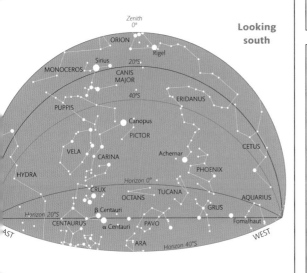

Looking south

WHOLE-SKY CHARTS »

January features

☐ **M42 (Orion)**. This diffuse nebula forms part of the sword that hangs from Orion's belt. It is one of the most celebrated objects in the entire sky and the most prominent of all nebulae, being visible to the naked eye under good conditions as a hazy, milky white or greenish patch, although it is better seen with binoculars. A huge cloud of dust and gas, M42 consists mainly of hydrogen gas, which shows as a reddish colour on photographs. It glows by the light of a star that formed within it, Theta-1 (θ¹) Orionis, also called the Trapezium since it in fact consists of four stars, visible through a small telescope. See also pp.114–15.

Also visible
- **Castor** (p.151)
- ⁝⁝ **M36, M37, and M38** (p.211)
- ⁝⁝ **M41** (p.151)
- ⁝⁝ **M44** (p.157)
- ☐ **NGC 2244** (p.151)
- **The Hyades and Pleiades** (p.211)

January meteors

The Quadrantids. This shower, visible to northern-hemisphere observers only, appears in the first week of January, radiating from northern Boötes, near the handle of the Plough. This area was once known as the constellation of Quadrans, hence the shower's name. Activity reaches about 100 meteors an hour on 3 and 4 January, but the peak is short-lived, the meteors are usually faint, and the radiant does not rise very high until after midnight.

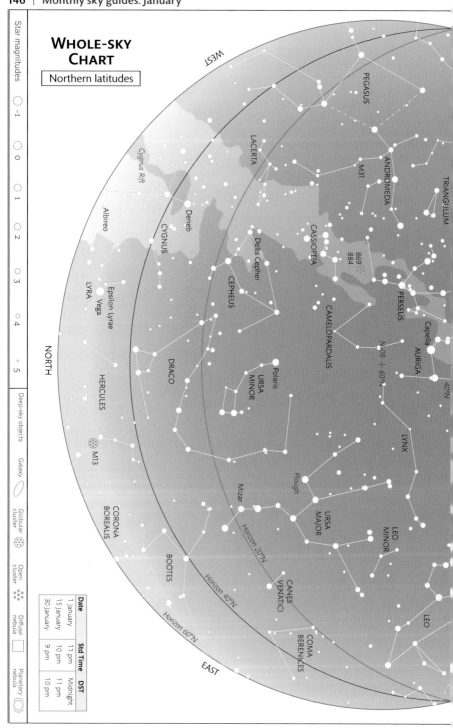

WHOLE-SKY CHART

Northern latitudes

Star magnitudes

○ -1
○ 0
○ 1
○ 2
○ 3
○ 4
○ 5

Deep-sky objects

Galaxy ◯
Globular cluster ⊛
Open cluster ⁝⁝⁝
Diffuse nebula ☐
Planetary nebula ◎

NORTH

WEST

EAST

PEGASUS
LACERTA
ANDROMEDA
M31
TRIANGULUM
Cygnus Rift
CASSIOPEIA
869
884
PERSEUS
Albireo
Deneb
CYGNUS
Delta Cephei
Capella
AURIGA
40°N
LYRA
Epsilon Lyrae
Vega
CEPHEUS
CAMELOPARDALIS
N.09 + 60°N
LYNX
DRACO
Polaris
URSA MINOR
HERCULES
⊛ M13
Mizar
Plough
URSA MAJOR
LEO MINOR
CORONA BOREALIS
Horizon 20°N
CANES VENATICI
LEO
BOÖTES
Horizon 40°N
COMA BERENICES
Horizon 60°N

Date	Std Time	DST
1 January	11 pm	Midnight
15 January	10 pm	11 pm
30 January	9 pm	10 pm

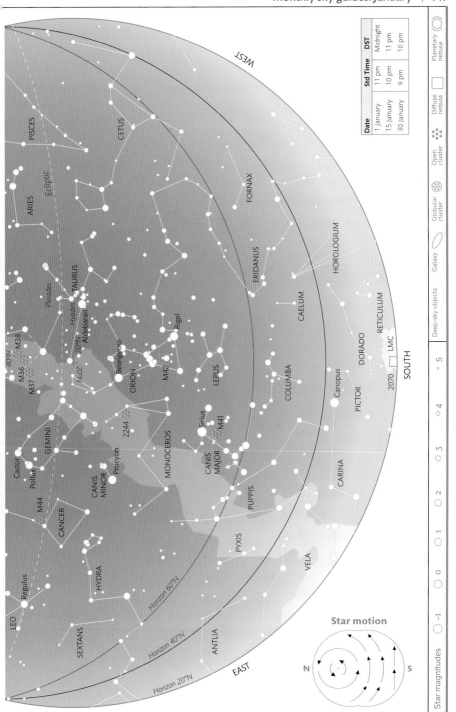

Date	Std Time	DST
1 January	11 pm	Midnight
15 January	10 pm	11 pm
30 January	9 pm	10 pm

Deep-sky objects

Galaxy
Globular cluster
Open cluster
Diffuse nebula
Planetary nebula

Star magnitudes

-1 0 1 2 3 4 5

Star motion

N   S

WEST

SOUTH

EAST

PISCES
CETUS
ARIES
Ecliptic
FORNAX
Pleiades
TAURUS
Hyades
ERIDANUS
HOROLOGIUM
Aldebaran
20°N
CAELUM
RETICULUM
40°N
M36
M38
NGC
ORION
Betelgeuse
Rigel
DORADO
M37
M42
LEPUS
COLUMBA
PICTOR
Canopus
LMC
2070
GEMINI
Castor
Pollux
2244
Sirius
M41
CANIS MAJOR
CARINA
M44
Procyon
CANIS MINOR
MONOCEROS
PUPPIS
CANCER
PYXIS
VELA
HYDRA
SEXTANS
ANTLIA
Regulus
LEO

Horizon 60°N
Horizon 40°N
Horizon 20°N

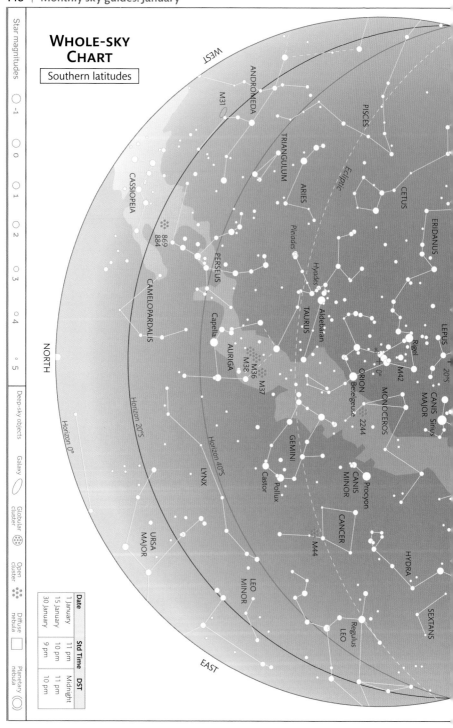

WHOLE-SKY CHART

Southern latitudes

Star magnitudes

○ -1
○ 0
○ 1
○ 2
○ 3
∘ 4
∘ 5

Deep-sky objects

Galaxy

Globular cluster

Open cluster

Diffuse nebula

Planetary nebula

WEST

NORTH

EAST

ANDROMEDA
M31
TRIANGULUM
CASSIOPEIA
869
884
PERSEUS
CAMELOPARDALIS
Capella
AURIGA
M38
M36
M37
LYNX
URSA MAJOR
LEO MINOR
PISCES
ARIES
Pleiades
Hyades
TAURUS
Aldebaran
GEMINI
Castor
Pollux
CANIS MINOR
CANCER
M44
LEO
Regulus
CETUS
ERIDANUS
Ecliptic
0°
20°S
LEPUS
Rigel
M42
ORION
Betelgeuse
MONOCEROS
2244
CANIS MAJOR
Sirius
Procyon
HYDRA
SEXTANS

Horizon 0°
Horizon 20°S
Horizon 40°S

Date	Std Time	DST
1 January	11 pm	Midnight
15 January	10 pm	11 pm
30 January	9 pm	10 pm

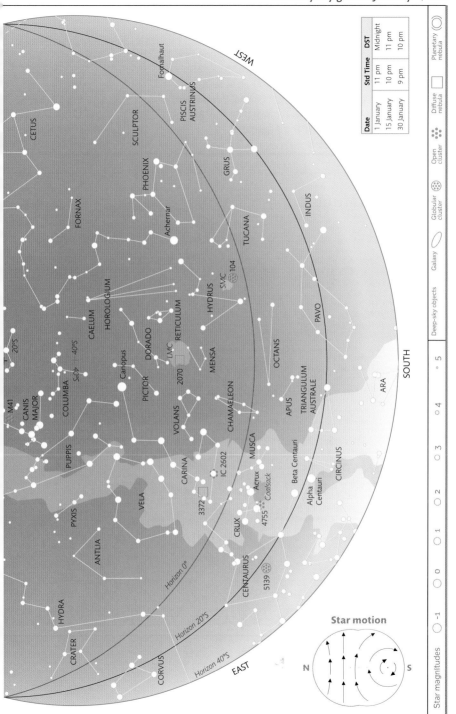

Date	Std Time	DST
1 January	11 pm	Midnight
15 January	10 pm	11 pm
30 January	9 pm	10 pm

WEST

SOUTH

EAST

CETUS
Fomalhaut
PISCIS AUSTRINUS
SCULPTOR
PHOENIX
FORNAX
Achernar
GRUS
INDUS
HOROLOGIUM
CAELUM
DORADO
RETICULUM
LMC
HYDRUS
MENSA
SMC
104
TUCANA
PAVO
PICTOR
Canopus
VOLANS
CHAMAELEON
OCTANS
APUS
TRIANGULUM AUSTRALE
ARA
20°S
40°S
40°S
COLUMBA
CANIS MAJOR
M41
PUPPIS
CARINA
IC 2602
MUSCA
2070
Beta Centauri
CIRCINUS
VELA
PYXIS
3372
Acrux
Coalsack
4755
CRUX
Alpha Centauri
ANTLIA
5139
CENTAURUS
Horizon 0°
HYDRA
CRATER
Horizon 20°S
CORVUS
Horizon 40°S

Star motion

N
S

Deep-sky objects

Galaxy · Globular cluster · Open cluster · Diffuse nebula · Planetary nebula

Star magnitudes

-1 · 0 · 1 · 2 · 3 · 4 · 5

FEBRUARY

SIRIUS, the brightest star, is visible in the evening sky at all latitudes, while observers in the south can also see Canopus, the second-brightest star. High in southern skies are the three constellations – Carina, Puppis, and Vela – that the ancient Greeks regarded as one: Argo Navis, the ship of Jason and the Argonauts.

SUNRISE AND SUNSET ON 15 FEBRUARY

Latitude	Sunrise	Sunset
60°N	07.40	16.50
40°N	06.50	17.40
20°N	06.30	18.00
0°	06.10	18.20
20°S	05.50	18.40
40°S	05.30	19.00

1 February: Std time 🕐 DST 🕐	15 February: Std time 🕐 DST 🕐	1 March: Std time 🕐 DST 🕐

NORTHERN LATITUDES

LOOKING NORTH Perseus and Cassiopeia are easy to find in the northwest, with the bright star Capella in Auriga above them. The familiar seven-star outline of the Plough (or Big Dipper) is the main feature in the northeast. Due north, the faint constellations Cepheus and Draco skim the horizon.

LOOKING SOUTH Orion and Taurus are prominent in the southwest. Sirius, the brightest star in the sky, lies almost due south, with Procyon above and slightly to its left. Castor and Pollux, the two brightest stars in Gemini, are overhead, with Cancer to their left. In the southeast, the distinctive shape of Leo is becoming well placed.

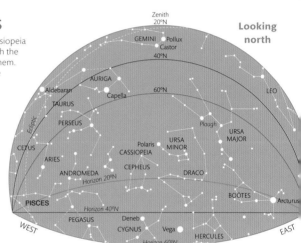

SOUTHERN LATITUDES

LOOKING NORTH Castor and Pollux in Gemini lie due north, with Procyon in Canis Minor above them. Sirius shines near the zenith. The stars of Orion and Taurus are prominently placed in the northwest, while Leo follows Cancer into the northeastern sky.

LOOKING SOUTH The three parts of Argo Navis, the ship of the Argonauts, are easily seen: Carina, the keel (with its brightest star Canopus); Puppis, the stern; and Vela, the sails, in between them. Achernar, in Eridanus, lies in the southwest with the Magellanic Clouds close by (see p.155). In the southeast, Centaurus and Crux (the Southern Cross) are rising.

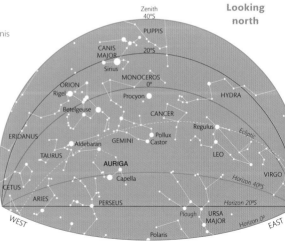

The Winter Triangle
Three brilliant stars in separate constellations form an equilateral triangle in the skies of northern winter and southern summer. Sirius in Canis Major forms the southern apex, while Betelgeuse (top right) in Orion and Procyon (top left) in Canis Minor complete the figure. The Winter Triangle straddles the celestial equator, and is clearly visible from all latitudes.

Star magnitudes ◯ -1   ◯ 0   ◯ 1   ∘ 2   · 3

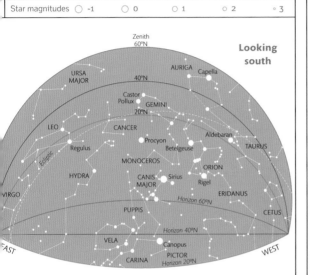

Zenith 60°N

Looking south

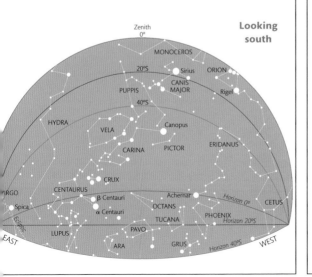

Zenith 0°

Looking south

WHOLE-SKY CHARTS »

February features
• **Castor (Gemini).** This star, also known as Alpha (α) Geminorum, can be seen as a close double through a telescope with magnification of about 100 times. The two stars orbit each other about every 450 years. There are other stars connected to Castor by gravity, too, completing a family of six stars in all, including a pair of faint red dwarfs. All six stars formed from the same cloud of gas. See also p.96.

✷ **M41 (Canis Major).** Observers in high northern latitudes often overlook this open cluster four degrees south of Sirius because from their locations it lies close to the horizon. However, it is easy to see with binoculars and can be detected with the naked eye under favourable conditions. See also p.74.

✷ **NGC 2244 (Monoceros).** The brightest members of this large open cluster are easily visible through binoculars. The stars are arranged in a rectangle, but the brightest of them, 6th-magnitude 12 Monocerotis, is a foreground object. The cluster lies at the centre of the Rosette Nebula (NGC 2237), a flower-like loop of gas that shows up well only on photographs. See also pp.110–11.

Also visible
✷ **M36, M37, and M38** (p.211)
☐ **M42** (p.145)
✷ **M44** (p.157)
☐ **NGC 3372** (p.151)
✷ **IC 2602** (p.157)
• **The Hyades and Pleiades** (p.211)

WHOLE-SKY CHART

Northern latitudes

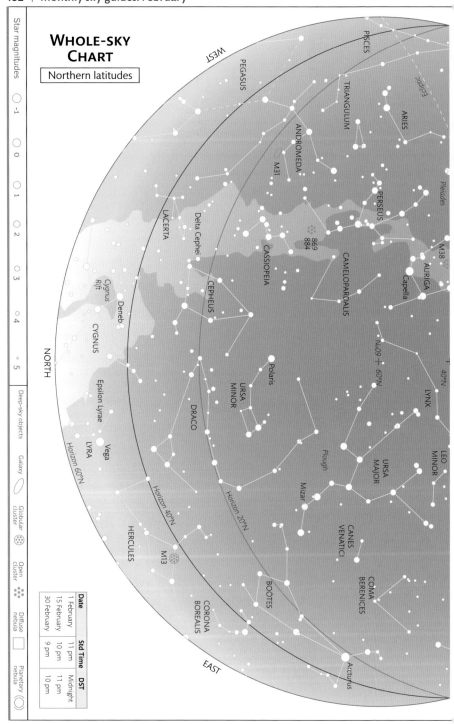

NORTH

WEST

EAST

Star magnitudes

○	-1
○	0
○	1
○	2
○	3
○	4
°	5

Deep-sky objects

Galaxy ⬭

Globular cluster ⬡

Open cluster ⁘

Diffuse nebula ▢

Planetary nebula ◎

Date	Std Time	DST
1 February	11 pm	Midnight
15 February	10 pm	11 pm
30 February	9 pm	10 pm

PISCES
PEGASUS
TRIANGULUM
ARIES
Ecliptic
ANDROMEDA
M31
Pleiades
PERSEUS
M38
AURIGA
Capella
LACERTA
CASSIOPEIA
869
884
CAMELOPARDALIS
T
40°N
Delta Cephei
CEPHEUS
60°N
LYNX
Cygnus Rift
Polaris
LEO MINOR
Deneb
URSA MINOR
CYGNUS
Plough
URSA MAJOR
DRACO
Epsilon Lyrae
Mizar
Horizon 60°N
LYRA
Vega
CANES VENATICI
COMA BERENICES
Horizon 40°N
Horizon 20°N
HERCULES
M13
BOÖTES
CORONA BOREALIS
Arcturus

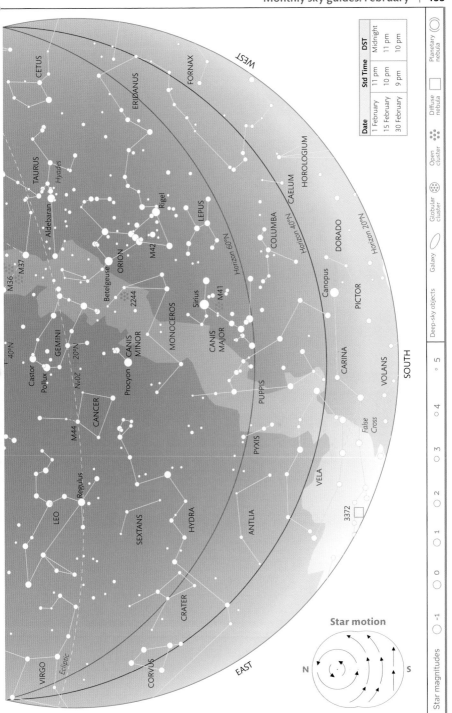

Date	Std Time	DST
1 February	11 pm	Midnight
15 February	10 pm	11 pm
30 February	9 pm	10 pm

Star magnitudes ◯ -1 ◯ 0 ◯ 1 ○ 2 ○ 3 ○ 4 ∘ 5

Deep-sky objects Galaxy Globular cluster Open cluster Diffuse nebula Planetary nebula

Star motion

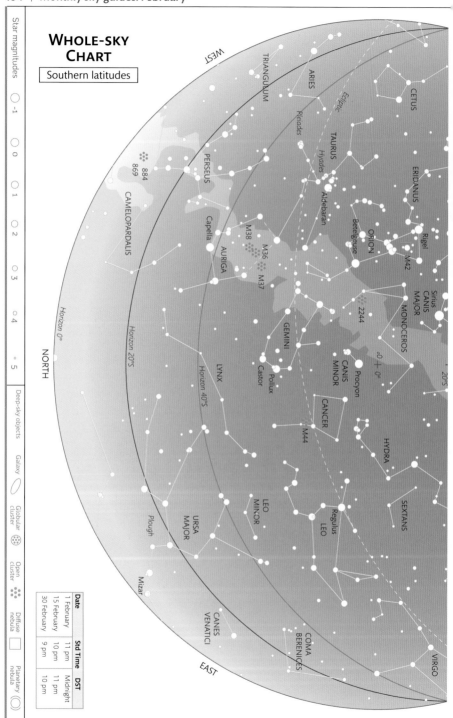

WHOLE-SKY CHART

Southern latitudes

WEST

TRIANGULUM

ARIES

CETUS

ERIDANUS

Ecliptic

Pleiades

TAURUS

Hyades

Aldebaran

884
869

PERSEUS

CAMELOPARDALIS

Capella

AURIGA

M38

M36

M37

Betelgeuse

ORION

Rigel

M42

Sirius

CANIS
MAJOR

MONOCEROS

2244

0°
0°

GEMINI

Castor

Pollux

CANIS
MINOR

Procyon

LYNX

CANCER

M44

HYDRA

SEXTANS

Horizon 0°

Horizon 20°S

Horizon 40°S

NORTH

LEO
MINOR

LEO

Regulus

URSA
MAJOR

Plough

Mizar

CANES
VENATICI

COMA
BERENICES

VIRGO

EAST

20°S

Star magnitudes							
○	○	○	○	○	○	°	
-1	0	1	2	3	4	5	

Deep-sky objects				
Galaxy	Globular cluster	Open cluster	Diffuse nebula	Planetary nebula
○	☼	⁙	☐	◎

Date	Std Time	DST
1 February	11 pm	Midnight
15 February	10 pm	11 pm
30 February	9 pm	10 pm

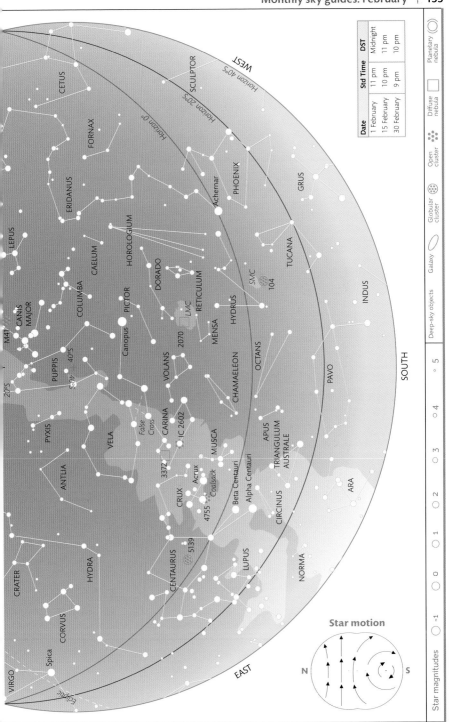

Date	Std Time	DST
1 February	11 pm	Midnight
15 February	10 pm	11 pm
30 February	9 pm	10 pm

WEST

SCULPTOR

CETUS

FORNAX

ERIDANUS

LEPUS

Horizon 0°S

Horizon 20°S

Horizon 40°S

PHOENIX

Achernar

GRUS

TUCANA

CANIS MAJOR

M47

CAELUM

COLUMBA

HOROLOGIUM

DORADO

PICTOR

RETICULUM

SMC

104

Canopus

VOLANS

MENSA

HYDRUS

INDUS

PUPPIS

40°S

2070

LMC

CHAMAELEON

OCTANS

PAVO

PYXIS

CARINA

IC 2602

MUSCA

APUS

ANTLIA

VELA

False Cross

337Z

Acrux

Coalsack

TRIANGULUM AUSTRALE

CIRCINUS

SOUTH

CRUX

4755

Beta Centauri

Alpha Centauri

ARA

HYDRA

5139

CENTAURUS

LUPUS

NORMA

CRATER

CORVUS

VIRGO

Spica

Ecliptic

EAST

Star motion

N   S

SMC 20°S

Deep-sky objects

° 5   ○ 4   ○ 3   ○ 2   ○ 1   ○ 0   ○ -1

Galaxy ◯   Globular cluster ◉   Open cluster ⁑   Diffuse nebula ▦   Planetary nebula ◎

Star magnitudes

MARCH

DAY AND NIGHT become nearly equal in length
the world over as the equinox approaches, around
21 March, the start of northern spring and southern
autumn. Orion and the bright stars around it are in the
west, while from southern latitudes rich fields of stars,
from Carina to Centaurus, lie south and southeast.

SUNRISE AND SUNSET ON 15 MARCH

Latitude	Sunrise	Sunset
60°N	06.20	18.00
40°N	06.10	18.10
20°N	06.10	18.10
0°	06.10	18.10
20°S	06.00	18.20
40°S	06.00	18.20

1 March: Std time 🕐 DST 🕐	15 March: Std time 🕐 DST 🕐	30 March: Std time 🕐 DST 🕐

NORTHERN LATITUDES

LOOKING NORTH The Plough
(or Big Dipper) stands high in the
northeast. Its bowl opens downwards,
towards Polaris, while its handle
points east to Arcturus, whose rising
signals the arrival of spring. Capella
is the most prominent star in the
northwest, with Perseus and
Cassiopeia now sinking low
towards the western horizon.
LOOKING SOUTH Leo, shaped
like a crouching lion, lies due
south, with the fainter stars of
Cancer to its right. Virgo, with
its brightest star Spica, is rising
in the southeast, while Orion
and the other stars of winter
depart in the southwest. Sirius
appears to twinkle above the
southwestern horizon.

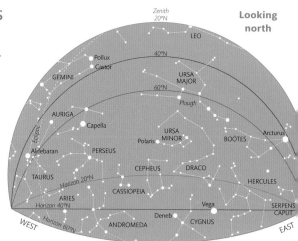

Looking north

SOUTHERN LATITUDES

LOOKING NORTH The zodiacal
constellations of Gemini, Cancer,
and Leo follow each other across
the sky from northwest to northeast.
The faint stars around the head
of Hydra are high in the north.
Arcturus can just be seen above
the northeastern horizon, heralding
the change of season.
LOOKING SOUTH Carina and
Vela, which lie almost due south,
are densely packed with stars.
Sirius, Canopus, and the Large
Magellanic Cloud (see p.161)
lie to the southwest. In the
southeast, Centaurus and Crux
(the Southern Cross) are rising.
Orion is sinking in the west,
while Virgo and the other stars
of winter enter from the east.

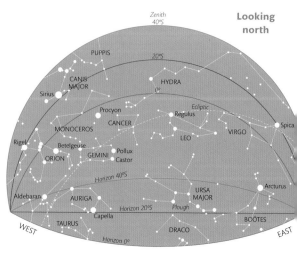

Looking north

The False Cross
Four stars in the adjacent constellations Carina and Vela form the False Cross, sometimes mistaken for the smaller but brighter true Southern Cross. The stars are Kappa (κ) and Delta (δ) Velorum (top left and right) and Iota (ι) and Epsilon (ε) Carinae (bottom left and right).

WHOLE-SKY CHARTS ⟩⟩

March features

✣ **M44 (Cancer).** Cancer, the faintest constellation of the zodiac, contains a major attraction: the open cluster M44, popularly known as Praesepe or the Beehive. Binoculars show it as a rounded patch of faint stars, resembling a swarm of bees. It can be detected with the naked eye in country skies. The stars Gamma (γ) and Delta (δ) Cancri, to the north and south of the cluster, were visualized in ancient times as donkeys feeding at a manger. See also p.72.

☐ **NGC 3372 (Carina).** This diffuse nebula, which contains the 6th-magnitude star Eta (η) Carinae, is commonly known as the Eta Carinae Nebula. On dark nights it can be seen with the naked eye, appearing as a particularly bright patch in the Milky Way. Its large size makes it ideal for observation through binoculars. A dark, V-shaped lane of dust runs through it. See also pp.76–77.

✣ **IC 2602 (Carina).** This bright open star cluster, popularly termed the Southern Pleiades, consists of a handful of naked-eye stars, the brightest of which is Theta (θ) Carinae, magnitude 2.8. Binoculars show about two dozen stars. See also pp.76–77.

Star magnitudes ◯ –1   ◯ 0   ◯ 1   ◦ 2   · 3

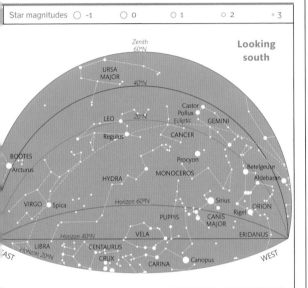

Zenith 60°N
Looking south

URSA MAJOR
40°N
Castor
Pollux
Ecliptic
LEO
20°N
GEMINI
Regulus
CANCER
BOÖTES
Arcturus
Procyon
MONOCEROS
Betelgeuse
Aldebaran
HYDRA
VIRGO Spica
Horizon 60°N
Sirius
ORION
PUPPIS
CANIS MAJOR
Rigel
LIBRA
Horizon 40°N
CENTAURUS
VELA
ERIDANUS
Horizon 20°N
CRUX
CARINA
Canopus
WEST
AST

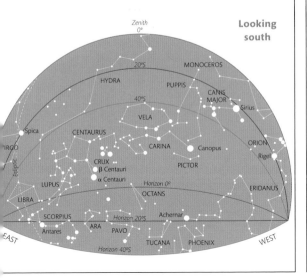

Zenith 0°
Looking south

20°S
MONOCEROS
HYDRA
PUPPIS
40°S
CANIS MAJOR
Sirius
VELA
CENTAURUS
CARINA
Canopus
ORION
Rigel
IRGO
CRUX
PICTOR
Spica
β Centauri
α Centauri
Horizon 0°
Ecliptic
LUPUS
OCTANS
ERIDANUS
LIBRA
Horizon 20°S
Achernar
SCORPIUS
Antares
ARA
PAVO
TUCANA
PHOENIX
WEST
EAST
Horizon 40°S

Also visible

● **Acrux** (P.163)
● **Castor** (P.151)
✣ **M36, M37, and M38** (P.211)
✣ **M41** (P.151)
☐ **M42** (P.145)
☐ **NGC** 2244 (P.151)
✣ **NGC** 4755 (P.163)

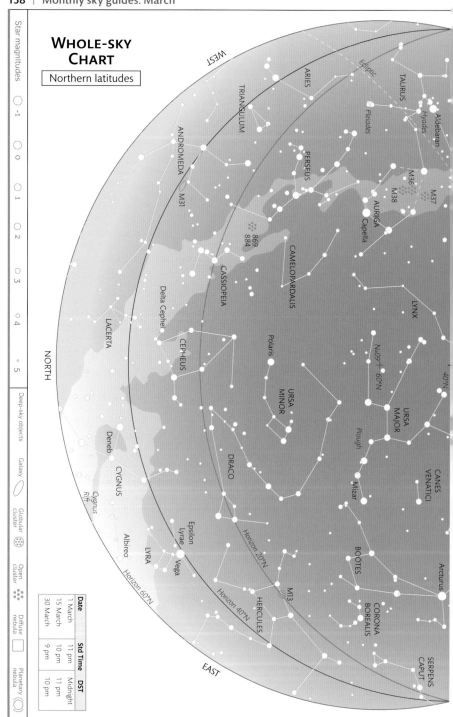

WHOLE-SKY CHART

Northern latitudes

Star magnitudes

○ -1
○ 0
○ 1
○ 2
○ 3
○ 4
○ 5

Deep-sky objects

Galaxy ⬭

Globular cluster ⬤

Open cluster ⁙

Diffuse nebula ▢

Planetary nebula ◎

WEST

ARIES

TRIANGULUM

ANDROMEDA

M31

Ecliptic

TAURUS

Pleiades

Hyades

Aldebaran

M36
M38

M37

PERSEUS

AURIGA

Capella

869
884

CAMELOPARDALIS

LYNX

CASSIOPEIA

Delta Cephei

CEPHEUS

Polaris

60°N
+90°N

URSA MINOR

URSA MAJOR

40°N

CANES VENATICI

LACERTA

DRACO

Plough

Mizar

BOÖTES

Deneb

CYGNUS

Cygnus Rift

Epsilon Lyrae

Vega

LYRA

Albireo

Horizon 20°N

M13

HERCULES

CORONA BOREALIS

Arcturus

SERPENS CAPUT

NORTH

Horizon 40°N

Horizon 60°N

EAST

Date	Std Time	DST
1 March	11 pm	Midnight
15 March	10 pm	11 pm
30 March	9 pm	10 pm

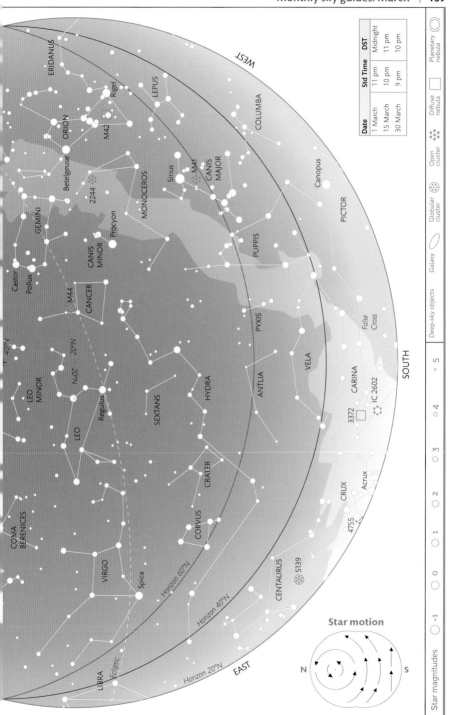

Date	Std Time	DST
1 March	11 pm	Midnight
15 March	10 pm	11 pm
30 March	9 pm	10 pm

WEST

SOUTH

EAST

Star motion

Star magnitudes

−1	0	1	2	3	4	5	

Deep-sky objects   Galaxy   Globular cluster   Open cluster   Diffuse nebula   Planetary nebula

WHOLE-SKY CHART

Southern latitudes

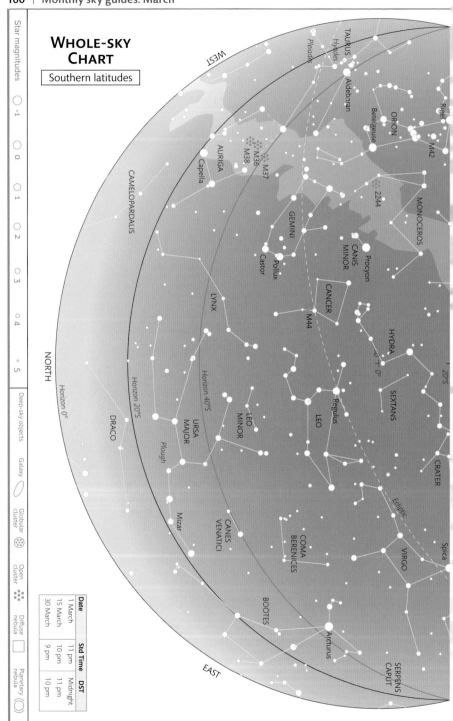

Star magnitudes

-1
0
1
2
3
4
5

Deep-sky objects

Galaxy

Globular cluster

Open cluster

Diffuse nebula

Planetary nebula

Date	Std Time	DST
1 March	11 pm	Midnight
15 March	10 pm	11 pm
30 March	9 pm	10 pm

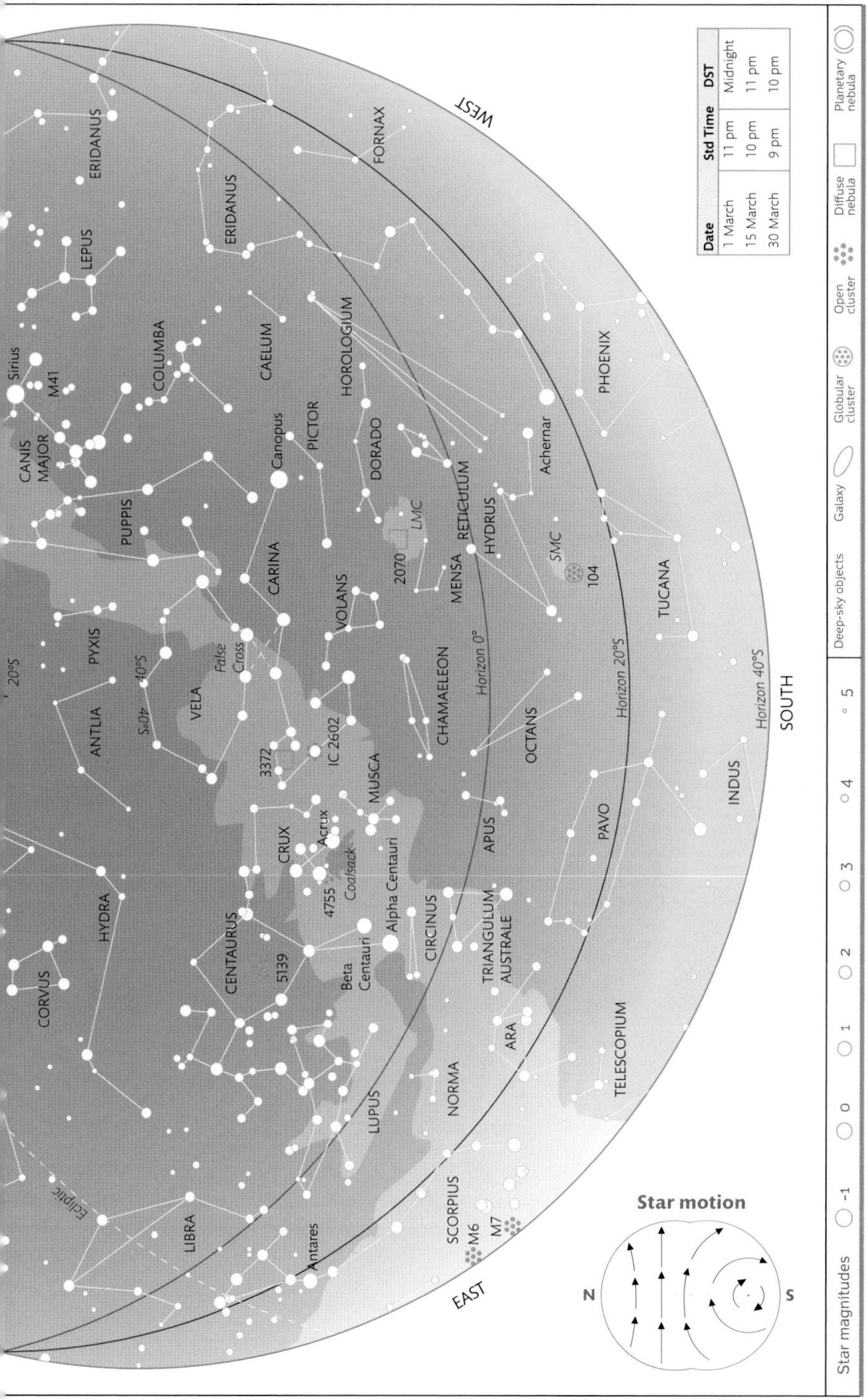

Date	Std Time	DST
1 March	11 pm	Midnight
15 March	10 pm	11 pm
30 March	9 pm	10 pm

WEST

SOUTH

EAST

Star motion

N — S

Star magnitudes
○ -1 ○ 0 ○ 1 ○ 2 ○ 3 ○ 4 ○ 5

Deep-sky objects

Galaxy ◯   Globular cluster ⊛   Open cluster ⁘   Diffuse nebula ▢   Planetary nebula ◉

ERIDANUS
LEPUS
Sirius
M41
CANIS MAJOR
COLUMBA
PUPPIS
CAELUM
PICTOR
HOROLOGIUM
FORNAX
DORADO
Canopus
CARINA
VOLANS
PYXIS
ANTLIA
VELA
False Cross
3372
IC 2602
MUSCA
CRUX
Acrux
4755
Coalsack
CENTAURUS
5139
Beta Centauri
Alpha Centauri
CIRCINUS
CHAMAELEON
APUS
MENSA
LMC
2070
RETICULUM
HYDRUS
Achernar
PHOENIX
SMC
104
TUCANA
OCTANS
INDUS
PAVO
TRIANGULUM AUSTRALE
ARA
NORMA
TELESCOPIUM
LUPUS
LIBRA
SCORPIUS
Antares
M6
M7
HYDRA
CORVUS
Ecliptic
20°S
40°S
Sco
Horizon 0°
Horizon 20°S
Horizon 40°S

APRIL

DAYLIGHT-SAVING TIME is in use in most northern latitudes, while darkness now arrives earlier in the south. The Plough (or Big Dipper) is high in the northern sky, Leo is prominent from all latitudes, and the Milky Way, containing a wealth of bright stars, is well placed for observers in the southern hemisphere.

SUNRISE AND SUNSET ON 15 APRIL

Latitude	Sunrise	Sunset
60°N	04.40	19.20
40°N	05.20	18.40
20°N	05.40	18.20
0°	06.00	18.00
20°S	06.10	17.50
40°S	06.30	17.30

1 April: Std time ⏲ DST ⏲	15 April: Std time ⏲ DST ⏲	30 April: Std time ⏲ DST ⏲

NORTHERN LATITUDES

LOOKING NORTH The familiar shape of the Plough (or Big Dipper) is almost overhead, the stars in its bowl pointing downwards to Polaris. Capella is in the northwest, while to its right the constellations Perseus and Cassiopeia sink low. In the east, Vega and Deneb lead the stars of summer into view.
LOOKING SOUTH Leo is high in the south. Beneath it is the relatively blank area of sky around Hydra. Left of Leo, in the southeast, the bright star Arcturus is prominently placed, with Spica in Virgo closer to the horizon. Above the western horizon, Castor and Pollux, in Gemini, are still visible, with Procyon just beneath them.

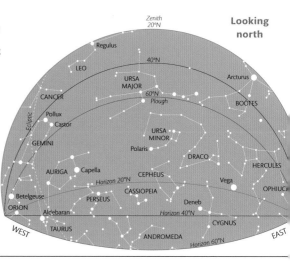

SOUTHERN LATITUDES

LOOKING NORTH Due north, Leo resembles a lion lying on its back, while Arcturus is prominent in the east, with Spica high above. Procyon remains on view in the northwest, as Castor and Pollux, in Gemini, set beneath it. The large but faint constellation of Hydra arches overhead from northwest to northeast.
LOOKING SOUTH Crux (the Southern Cross) is high in the south with Vela and Carina to its right and Alpha (α) and Beta (β) Centauri to its left. Canopus is in the southwest, the Large Magellanic Cloud (LMC) to its lower left (see p.167). Sirius is almost due west, while Scorpius lies southeast.

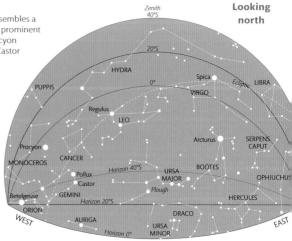

The Coalsack (Crux)
What appears to be a hole in the Milky Way next to Crux (the Southern Cross) is in fact a huge cloud of dust that blots out light from the stars behind it. This dark nebula, known as the Coalsack, is about 60 light years in diameter, and is thought to lie about 600 light years from us.

Star magnitudes ○ -1   ○ 0   ○ 1   ∘ 2   · 3

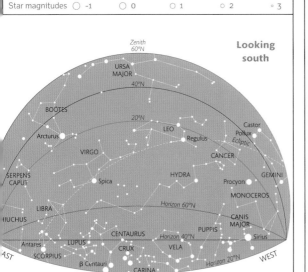

Zenith
60°N

Looking south

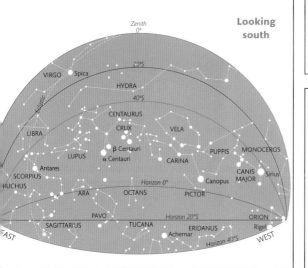

Zenith
0°

Looking south

WHOLE-SKY CHARTS »

April features

● **Acrux (Crux)**. The brightest star in the Southern Cross, known as Acrux or Alpha (α) Crucis, is in fact a close double star. A small telescope will divide it into two sparkling points. A fainter star, which is not related to the main pair, can be seen through binoculars. See also pp.86–87.

❖ **NGC 4755 (Crux)**. This open cluster, popularly known as the Jewel Box, is visible to the naked eye as a hazy star, and is best viewed through binoculars or a small telescope. The English astronomer Sir John Herschel described it in the 1830s as resembling "a casket of variously coloured precious stones". Although it appears to lie next to the Coalsack (see photograph, above), it is in fact over 12 times more distant. See also pp.86–87.

Also visible

● **Alpha (α) Centauri** (p.169)
● **Mizar** (p.169)
❖ **M44** (p.157)
❖ **NGC 2244** (p.151)
☐ **NGC 3372** (p.157)
❖ **NGC 5139** (p.157)
❖ **IC 2602** (p.157)

April meteors

The Lyrids. The Lyrid meteor shower, most easily seen from northern latitudes, reaches its peak around 21 April. It is one of the poorer showers, with a maximum rate of only about 10 meteors an hour, but the meteors are fast, bright, and often leave trains. The radiant is in the constellation Lyra, close to the bright star Vega.

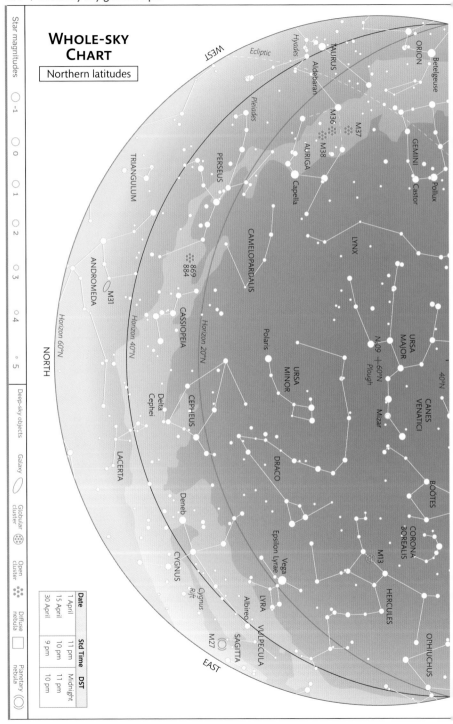

Star magnitudes

○ -1

○ 0

○ 1

○ 2

○ 3

○ 4

○ 5

Deep-sky objects

Galaxy ⬭

Globular cluster ⊛

Open cluster ⁛

Diffuse nebula ▫

Planetary nebula ◎

WHOLE-SKY CHART

Northern latitudes

NORTH

WEST

EAST

Ecliptic

Hyades

TAURUS

Aldebaran

Pleiades

PERSEUS

TRIANGULUM

ANDROMEDA

M31

Horizon 60°N

Horizon 40°N

Horizon 20°N

869
884

CAMELOPARDALIS

M36

M38

M37

AURIGA

Capella

ORION

Betelgeuse

GEMINI

Castor

Pollux

LYNX

URSA MAJOR

N09 +60°N

Plough

Mizar

CANES VENATICI

40°N

BOÖTES

CORONA BOREALIS

HERCULES

M13

OPHIUCHUS

Polaris

URSA MINOR

DRACO

CEPHEUS

Delta Cephei

CASSIOPEIA

LACERTA

Deneb

CYGNUS

Cygnus Rift

Albireo

LYRA

VULPECULA

Epsilon Lyrae

Vega

SAGITTA

M27

Date	Std Time	DST
1 April	11 pm	Midnight
15 April	10 pm	11 pm
30 April	9 pm	10 pm

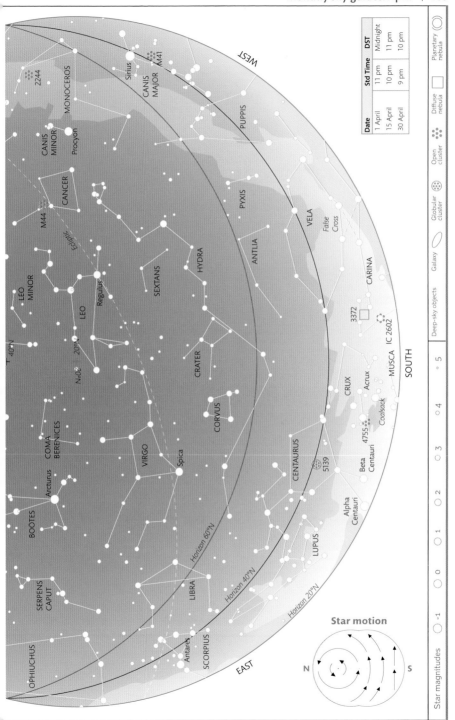

Date	Std Time	DST
1 April	11 pm	Midnight
15 April	10 pm	11 pm
30 April	9 pm	10 pm

Star motion

Star magnitudes ⬡ -1 ⬡ 0 ⬡ 1 ⬡ 2 ⬡ 3 ⬡ 4 ⬡ 5

Deep-sky objects — Galaxy — Globular cluster — Open cluster — Diffuse nebula — Planetary nebula

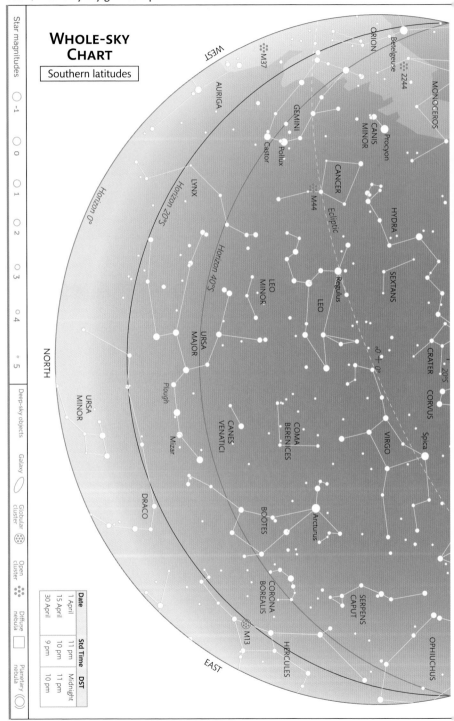

WHOLE-SKY CHART

Southern latitudes

Star magnitudes

○ -1
○ 0
○ 1
○ 2
○ 3
○ 4
° 5

Deep-sky objects

Galaxy ⬭
Globular cluster ⬡
Open cluster ⁙
Diffuse nebula ▢
Planetary nebula ◎

NORTH

WEST

EAST

Horizon 0°
Horizon 20°S
Horizon 40°S

Ecliptic

ORION
Betelgeuse
2244
MONOCEROS
M37
AURIGA
GEMINI
CANIS MINOR
Procyon
Castor
Pollux
CANCER
M44
HYDRA
LYNX
LEO MINOR
Regulus
LEO
SEXTANS
URSA MAJOR
URSA MINOR
Plough
Mizar
CANES VENATICI
COMA BERENICES
DRACO
BOÖTES
Arcturus
VIRGO
Spica
CRATER
CORVUS
CORONA BOREALIS
SERPENS CAPUT
M13
HERCULES
OPHIUCHUS

Date	Std Time	DST
1 April	11 pm	Midnight
15 April	10 pm	11 pm
30 April	9 pm	10 pm

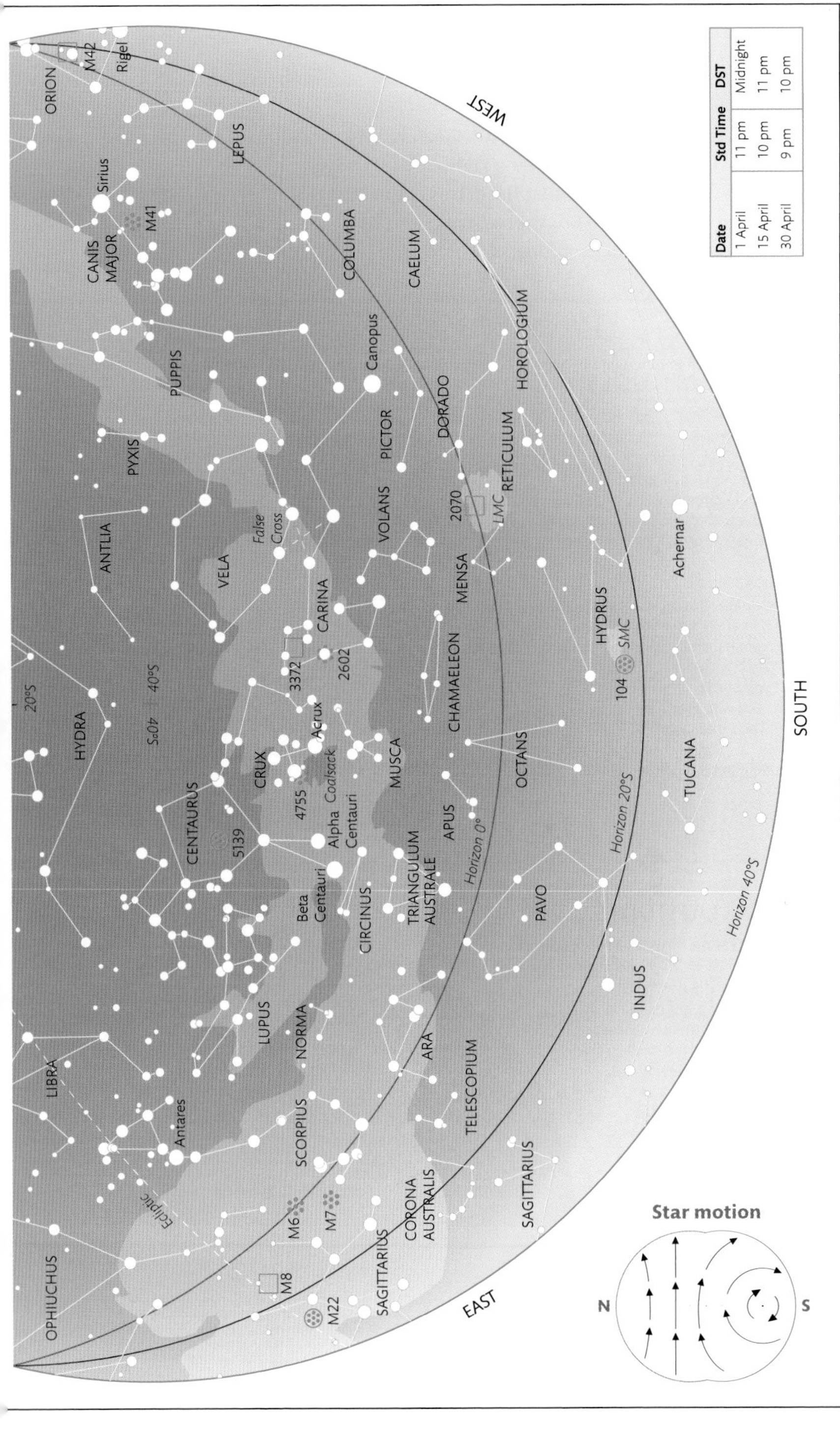

Date	Std Time	DST
1 April	11 pm	Midnight
15 April	10 pm	11 pm
30 April	9 pm	10 pm

WEST

ORION
M42
Rigel
LEPUS
M41
Sirius
CANIS MAJOR
COLUMBA
CAELUM
PUPPIS
Canopus
HOROLOGIUM
PYXIS
ANTLIA
PICTOR
DORADO
RETICULUM
VELA
False Cross
VOLANS
2070
LMC
MENSA
Achernar
CARINA
3372
2602
CHAMAELEON
HYDRUS
104
SMC
20°S
HYDRA
Sclr 40°S
Acrux
MUSCA
OCTANS
TUCANA
CRUX
CENTAURUS
4755
APUS
Horizon 0°
5139
Alpha Centauri
TRIANGULUM AUSTRALE
Horizon 20°S
Beta Centauri
CIRCINUS
PAVO
Horizon 40°S
INDUS
LUPUS
NORMA
ARA
LIBRA
TELESCOPIUM
Antares
SCORPIUS
CORONA AUSTRALIS
Ecliptic
M6
M7
M8
SAGITTARIUS
M22
OPHIUCHUS

SOUTH

EAST

Star motion

N
S

Star magnitudes

○ −1   ○ 0   ○ 1   ○ 2   ○ 3   ○ 4   ○ 5

Deep-sky objects

Galaxy   Globular cluster   Open cluster   Diffuse nebula   Planetary nebula

MAY

LENGTHENING TWILIGHT can make early evening observing more difficult in northern latitudes, where the sparkling stars of winter are being replaced by those of summer. In the southern hemisphere, this is the best time of year to see Crux (the Southern Cross) and its "pointers", Alpha and Beta Centauri.

SUNRISE AND SUNSET ON 15 MAY

Latitude	Sunrise	Sunset
60°N	03.20	20.30
40°N	04.40	19.10
20°N	05.20	18.30
0°	05.50	18.00
20°S	06.20	17.30
40°S	07.00	17.00

1 May: Std time ⏱ DST ⏱	15 May: Std time ⏱ DST ⏱	30 May: Std time ⏱ DST ⏱

NORTHERN LATITUDES

LOOKING NORTH The Plough (or Big Dipper) stands almost overhead, while Cygnus and the bright star Vega climb higher in the northeast as a foretaste of summer. The faint constellation of Cepheus is between Polaris and the northern horizon. Gemini is the last of the winter constellations to sink below the northwestern horizon.

LOOKING SOUTH The bright stars Arcturus in Boötes and Spica in Virgo are high in the south. Ophiuchus and both halves of Serpens occupy much of the southeastern sky, with Hercules above them. Leo dominates the southwest, with the largely blank area of sky occupied by Hydra beneath it.

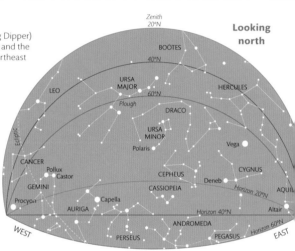

SOUTHERN LATITUDES

LOOKING NORTH Arcturus, in Boötes, glitters in the north, with Spica, in Virgo, nearly overhead. Hercules is close to the northeastern horizon, to the lower right of Arcturus, while Ophiuchus and the two halves of Serpens cover much of the eastern sky. Leo is prominent in the northwest.

LOOKING SOUTH Centaurus and its adjoining constellations dominate the southern sky, with Alpha (α) and Beta (β) Centauri pointing towards Crux (the Southern Cross). While Canopus sets in the southwest, the well-stocked constellations Sagittarius and Scorpius are rising in the south-east, a sign that winter is on its way.

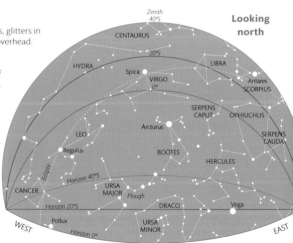

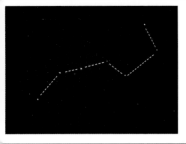

The Plough

The familiar shape of the Plough (or Big Dipper), formed by seven stars in Ursa Major (pp.136–37), is a prominent sight in northern skies in spring. The second star in the handle, called Mizar, has a fainter partner, visible to observers with normal eyesight (see panel, right).

Star magnitudes	○ -1	○ 0	○ 1	○ 2	° 3

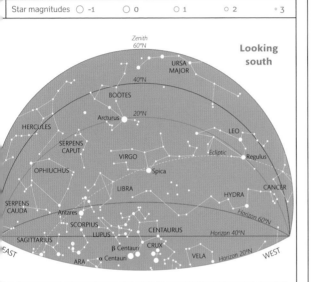

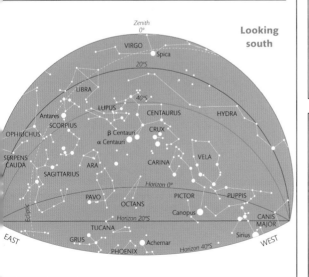

WHOLE-SKY CHARTS »

May features

• **Alpha (α) Centauri.** A mere 4.4 light years away, this is the closest star to the Earth after the Sun. A small telescope reveals two yellow stars, the brighter of which is similar in temperature and brightness to the Sun. A third member of the family, Proxima Centauri, is much fainter. See also p.79.

• **Mizar (Ursa Major).** The second star in the handle of the Plough, known as Mizar or Zeta (ζ) Ursae Majoris, is a wide double. Its companion, Alcor, is visible to the naked eye, while a small telescope shows another star, closer to Mizar. See also pp.136–37

⊛ **NGC 5139 (Centaurus).** The largest and brightest globular cluster in the sky, also called Omega (ω) Centauri, is visible to the naked eye as a hazy star. It is clearer through binoculars, and a small telescope shows the brightest of its 100,000 or more stars. See also p.79.

Also visible

- **Acrux** (p.163)
- **M6 and M7** (p.175)
- **M13** (p.175)
- **M44** (p.157)
- **NGC 4755** (p.163)

May meteors

The Eta Aquariids. These fast-moving, bright meteors, caused by dust from Halley's Comet, radiate from a point near the "water jar" in Aquarius, and are best seen from southern latitudes. Activity begins in late April, peaks at about 35 meteors an hour in the first week of May, and ends in late May.

WHOLE-SKY CHART

Northern latitudes

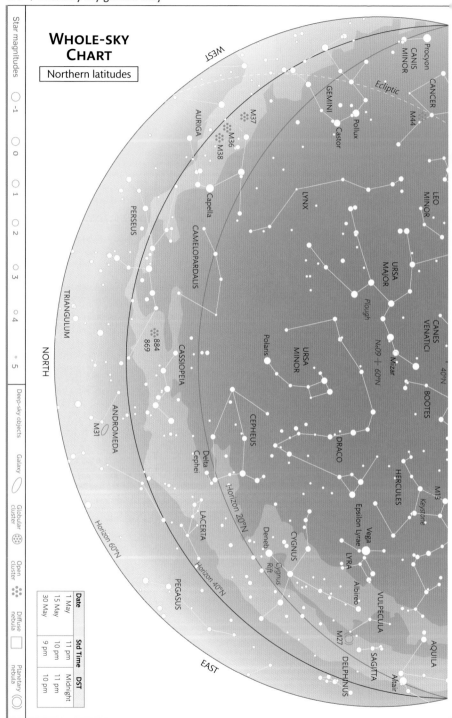

Star magnitudes

Star magnitudes:
- ○ -1
- ○ 0
- ○ 1
- ○ 2
- ○ 3
- ○ 4
- ° 5

Deep-sky objects

- Galaxy ⬭
- Globular cluster ⊛
- Open cluster ⁘
- Diffuse nebula ▢
- Planetary nebula ◎

NORTH

WEST

EAST

Ecliptic

CANIS MINOR
Procyon
CANCER
GEMINI
M44
Pollux
Castor
AURIGA
M37
M36
M38
Capella
LEO MINOR
LYNX
URSA MAJOR
Plough
CANES VENATICI
N 60°N
Mizar
N 60°N + 60°N
BOOTES
40°N
PERSEUS
CAMELOPARDALIS
TRIANGULUM
Polaris
URSA MINOR
884
869
CASSIOPEIA
CEPHEUS
DRACO
HERCULES
M13
Keystone
ANDROMEDA
M31
Delta Cephei
Epsilon Lyrae
Vega
LYRA
LACERTA
CYGNUS
Cygnus Rift
Deneb
Albireo
VULPECULA
PEGASUS
Horizon 20°N
Horizon 40°N
Horizon 60°N
M27
SAGITTA
DELPHINUS
AQUILA
Altair

Date	Std Time	DST
1 May	11 pm	Midnight
15 May	10 pm	11 pm
30 May	9 pm	10 pm

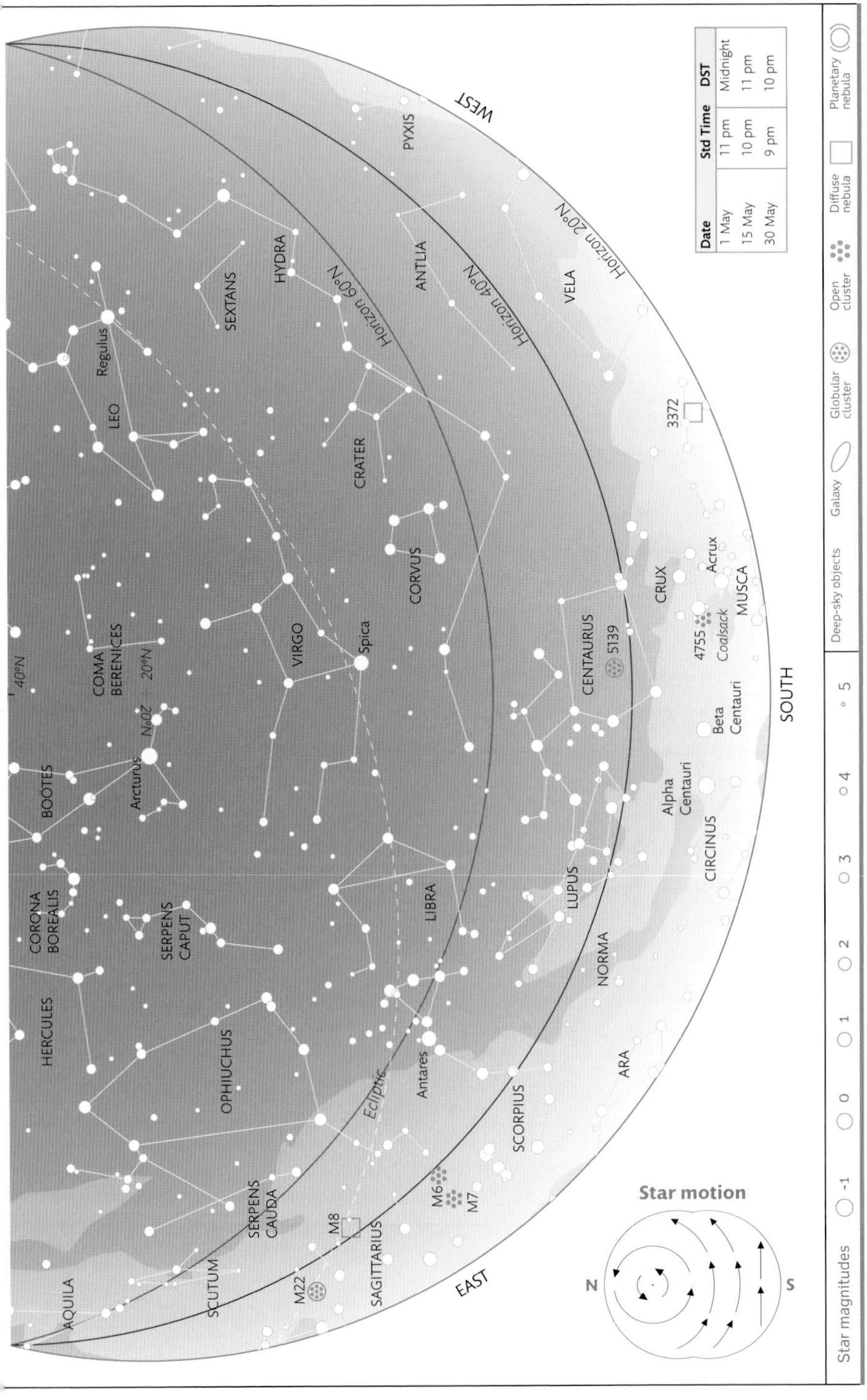

WEST

PYXIS

HYDRA

SEXTANS

ANTLIA

Regulus

VELA

LEO

Horizon 20°N

CRATER

Horizon 40°N

Horizon 60°N

CORVUS

3372

COMA
BERENICES

VIRGO

Spica

CRUX

Acrux

MUSCA

40°N

20°N

4755

Coalsack

5139

Arcturus

CENTAURUS

Beta
Centauri

BOOTES

SOUTH

CORONA
BOREALIS

SERPENS
CAPUT

LIBRA

LUPUS

Alpha
Centauri

CIRCINUS

HERCULES

NORMA

OPHIUCHUS

Ecliptic

Antares

ARA

SERPENS
CAUDA

SCORPIUS

M8

M6

SCUTUM

M7

M22

SAGITTARIUS

AQUILA

EAST

Star motion

N

S

Date	Std Time	DST
1 May	11 pm	Midnight
15 May	10 pm	11 pm
30 May	9 pm	10 pm

Deep-sky objects

| | Galaxy | Globular cluster | Open cluster | Diffuse nebula | Planetary nebula |

Star magnitudes

-1   0   1   2   3   4   5

WHOLE-SKY CHART

Southern latitudes

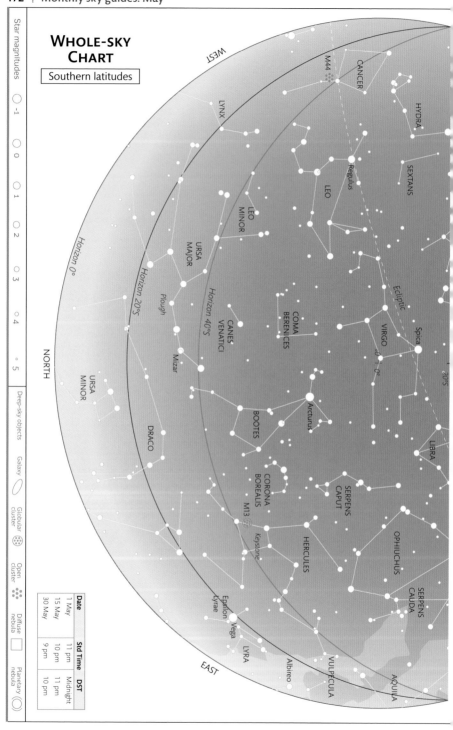

Star magnitudes

○ -1
○ 0
○ 1
○ 2
○ 3
○ 4
° 5

NORTH

Deep-sky objects

Galaxy ⬭

Globular cluster ⊛

Open cluster ⁙

Diffuse nebula ▢

Planetary nebula ◎

Date	Std Time	DST
1 May	11 pm	Midnight
15 May	10 pm	11 pm
30 May	9 pm	10 pm

WEST

M44

CANCER

HYDRA

LYNX

Regulus

SEXTANS

LEO

LEO MINOR

Horizon 0°

URSA MAJOR

COMA BERENICES

Ecliptic

Horizon 20°S

Plough

CANES VENATICI

Horizon 40°S

VIRGO

Spica

Mizar

URSA MINOR

BOOTES

Arcturus

20°S

LIBRA

DRACO

CORONA BOREALIS

SERPENS CAPUT

M13

OPHIUCHUS

Keystone

HERCULES

SERPENS CAUDA

Epsilon Lyrae

Vega

LYRA

Albireo

VULPECULA

AQUILA

EAST

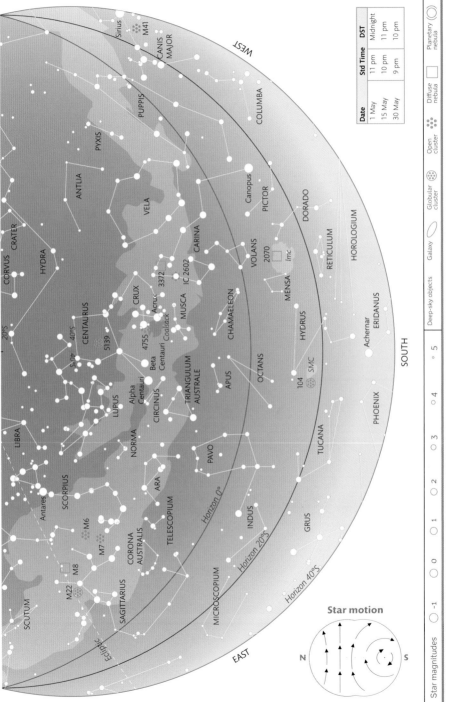

Date	Std Time	DST
1 May	11 pm	Midnight
15 May	10 pm	11 pm
30 May	9 pm	10 pm

WEST

SOUTH

EAST

Star motion

N   S

Star magnitudes ◯ -1 ◯ 0 ◯ 1 ◯ 2 ◯ 3 ◯ 4 ◦ 5

Deep-sky objects Galaxy Globular cluster Open cluster Diffuse nebula Planetary nebula

Sirius M41 CANIS MAJOR COLUMBA PUPPIS PYXIS ANTLIA VELA Canopus PICTOR DORADO HOROLOGIUM RETICULUM ERIDANUS Achernar HYDRUS VOLANS 2070 lmc MENSA CHAMAELEON CARINA IC 2602 MUSCA Coalsack Acrux CRUX 3372 4755 CENTAURUS 5139 40°S Beta Centauri Alpha Centauri CIRCINUS TRIANGULUM AUSTRALE APUS OCTANS SMC 104 TUCANA PHOENIX GRUS INDUS PAVO NORMA ARA TELESCOPIUM MICROSCOPIUM CORONA AUSTRALIS SAGITTARIUS M22 M8 SCUTUM M7 M6 SCORPIUS Antares LIBRA 20°S LUPUS HYDRA CRATER CORVUS

Horizon 0° Horizon 20°S Horizon 40°S Ecliptic

JUNE

NIGHTS ARE AT THEIR SHORTEST, and days at their longest in northern latitudes in June, and vice versa in the southern hemisphere. In the far north, twilight is now permanent throughout the night, while the winter skies of the south are dominated by a band of prominent constellations lying in the Milky Way.

SUNRISE AND SUNSET ON 15 JUNE

Latitude	Sunrise	Sunset
60°N	02.40	21.20
40°N	04.30	19.30
20°N	05.20	18.40
0°	06.00	18.00
20°S	06.30	17.30
40°S	07.20	16.40

1 June: Std time ⏱ DST ⏱	15 June: Std time ⏱ DST ⏱	30 June: Std time ⏱ DST ⏱

NORTHERN LATITUDES

• **LOOKING NORTH** Ursa Minor (the Little Bear) stands on its tail, and Draco arches above the north celestial pole, marked by Polaris. The Plough (or Big Dipper) is high in the northwest. Towards the east, the stars that form the Summer Triangle are in view: Vega, Deneb in Cygnus, and Altair in Aquila (far left of looking south map).

• **LOOKING SOUTH** Hercules is high in the southeast, with Vega to its left and Arcturus, in Boötes, to its right. Ophiuchus, Virgo, and both halves of Serpens are below them. Antares, in Scorpius, glints red above the southern horizon, the faint stars of Libra to its right. Leo is setting in the west.

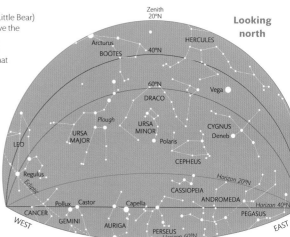

SOUTHERN LATITUDES

• **LOOKING NORTH** Boötes and Hercules are prominently placed almost due north, with Virgo, Ophiuchus, and both parts of Serpens above them. Antares (in Scorpius) is overhead. Altair (in Aquila) and Vega are the most prominent stars in the northeast. Leo is setting in the northwest.

• **LOOKING SOUTH** The constellations Sagittarius, Ara, Scorpius, Lupus, Centaurus, Crux (the Southern Cross), Carina, and Vela form a bright band, from southeast to southwest, against dense star fields in the Milky Way. The brightest stars along this band are Antares, Alpha (α) and Beta (β) Centauri, and Acrux (see also p.179).

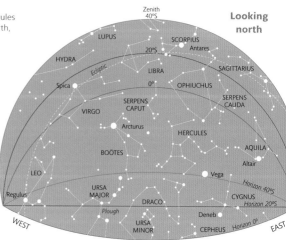

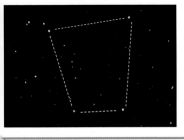

The Keystone (Hercules)

Four stars form a shape known as the Keystone that represents the pelvis of Hercules (pp.98–99). Clockwise from top left, they are Pi (π), Eta (η), Zeta (ζ), and Epsilon (ε) Herculis. The star cluster M13 (see panel, right) lies between Eta and Zeta.

WHOLE-SKY CHARTS »

June features

✣ **M6 (Scorpius).** This open cluster and the nearby M7 lie near the "sting" in the tail of Scorpius, the scorpion. M6 can be seen with the naked eye and is an impressive sight through binoculars. It is sometimes called the Butterfly Cluster because its stars are arranged in the shape of a butterfly with open wings. See also pp.126–27.

✣ **M7 (Scorpius).** This open cluster appears to the naked eye as a bright patch in front of dense fields of stars in the Milky Way, and is a superb sight when seen through binoculars. It is larger and brighter than its neighbour M6, being just over half the distance from us (980 light years, compared with 1,500 light years for M6). See also pp.126–27.

⊛ **M13 (Hercules).** This is the most prominent globular cluster in northern skies; even so, it is barely visible to the naked eye, and from urban areas it can be seen only with the aid of binoculars, through which it resembles a glowing ball with a bright centre. It lies on one side of the Keystone of Hercules (see photograph, above), one-third the way between Eta (η) and Zeta (ζ) Herculis. See also pp.98–99.

Star magnitudes ◯ -1 ◯ 0 ◯ 1 ◦ 2 ° 3

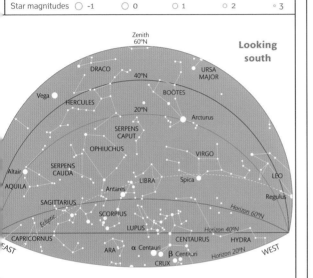

Zenith 60°N

Looking south

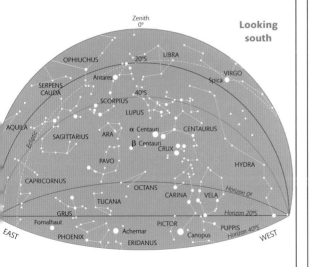

Zenith 0°

Looking south

Also visible

- **Albireo** (p.187)
- **Alpha (α) Centauri** (p.169)
- **Antares** (p.181)
- **Epsilon (ε) Lyrae** (p.181)
- **Mizar** (p.169)
- ☐ **M8** (p.181)
- ⊛ **M22** (p.181)
- ◯ **M27** (p.193)
- ⊛ **NGC 5139** (p.169)

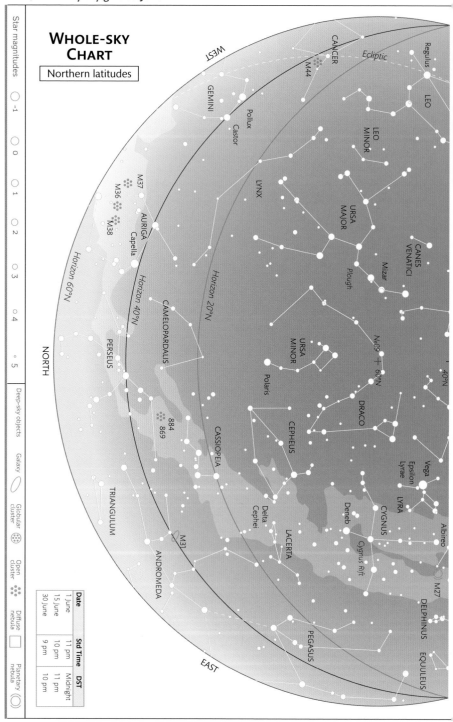

WHOLE-SKY CHART

Northern latitudes

Star magnitudes

○ -1
○ 0
○ 1
○ 2
○ 3
○ 4
° 5

Deep-sky objects

Galaxy ⬭
Globular cluster ⊛
Open cluster ⁙
Diffuse nebula ▢
Planetary nebula ◎

Date	Std Time	DST
1 June	11 pm	Midnight
15 June	10 pm	11 pm
30 June	9 pm	10 pm

NORTH

WEST

EAST

Ecliptic

CANCER
M44
GEMINI
Pollux
Castor
LYNX
M37
M36
M38
AURIGA
Capella
CAMELOPARDALIS
PERSEUS
Horizon 60°N
Horizon 40°N
Horizon 20°N
884
869
CASSIOPEIA
TRIANGULUM
M31
ANDROMEDA

Regulus
LEO
LEO MINOR
URSA MAJOR
CANES VENATICI
Mizar
Plough
URSA MINOR
Polaris
NGP
+60°N
DRACO
CEPHEUS
Delta Cephei
LACERTA
PEGASUS
EQUULEUS
DELPHINUS
M27
Albireo
CYGNUS
Cygnus Rift
Deneb
LYRA
Vega
Epsilon Lyrae
40°N
60°N

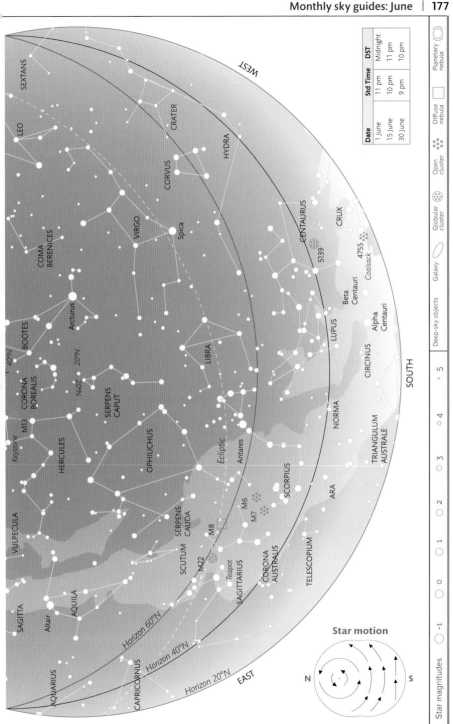

Star motion

Date	Std Time	DST
1 June	11 pm	Midnight
15 June	10 pm	11 pm
30 June	9 pm	10 pm

Deep-sky objects

Galaxy · Globular cluster · Open cluster · Diffuse nebula · Planetary nebula

Star magnitudes: -1, 0, 1, 2, 3, 4, 5

WEST

SOUTH

EAST

SEXTANS

LEO

CRATER

CORVUS

HYDRA

COMA BERENICES

VIRGO

Spica

CENTAURUS

5139

CRUX

4755

Coalsack

Arcturus

BOÖTES

Beta Centauri

Alpha Centauri

CORONA BOREALIS

SERPENS CAPUT

LIBRA

LUPUS

CIRCINUS

40°N

20°N

N 40°

M13

Keystone

HERCULES

OPHIUCHUS

Antares

Ecliptic

NORMA

TRIANGULUM AUSTRALE

VULPECULA

SERPENS CAUDA

SCUTUM

M8

M22

SCORPIUS

M6

M7

ARA

SAGITTA

AQUILA

Altair

Teapot

SAGITTARIUS

CORONA AUSTRALIS

TELESCOPIUM

AQUARIUS

CAPRICORNUS

Horizon 60°N

Horizon 40°N

Horizon 20°N

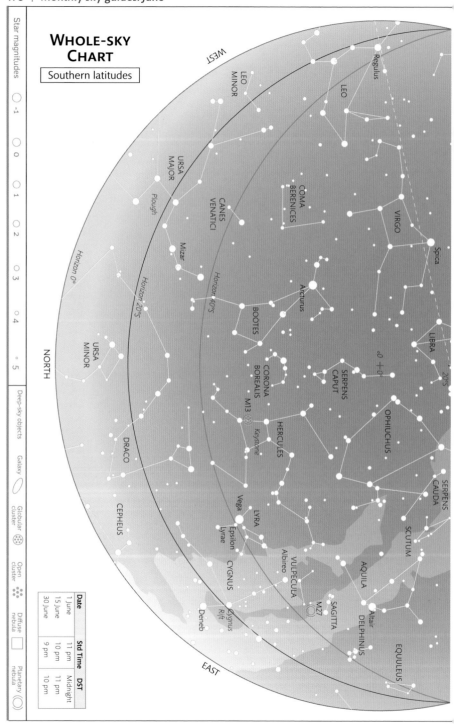

WHOLE-SKY CHART

Southern latitudes

WEST

NORTH

EAST

LEO MINOR

URSA MAJOR

Plough

Horizon 0°

Mizar

CANES VENATICI

Horizon 20°S

URSA MINOR

Horizon 40°S

BOÖTES

Arcturus

CORONA BOREALIS

M13

Keystone

HERCULES

DRACO

CEPHEUS

Vega

LYRA

Epsilon Lyrae

Albireo

CYGNUS

Cygnus Rift

Deneb

Regulus

LEO

COMA BERENICES

VIRGO

Spica

SERPENS CAPUT

0°

0°

LIBRA

20°S

OPHIUCHUS

SERPENS CAUDA

SCUTUM

VULPECULA

M27

SAGITTA

AQUILA

Altair

DELPHINUS

EQUULEUS

Star magnitudes

○ -1

○ 0

○ 1

○ 2

○ 3

○ 4

° 5

Deep-sky objects

Galaxy ⬭

Globular cluster ⊛

Open cluster ⸭

Diffuse nebula ▢

Planetary nebula ◉

Date	Std Time	DST
1 June	11 pm	Midnight
15 June	10 pm	11 pm
30 June	9 pm	10 pm

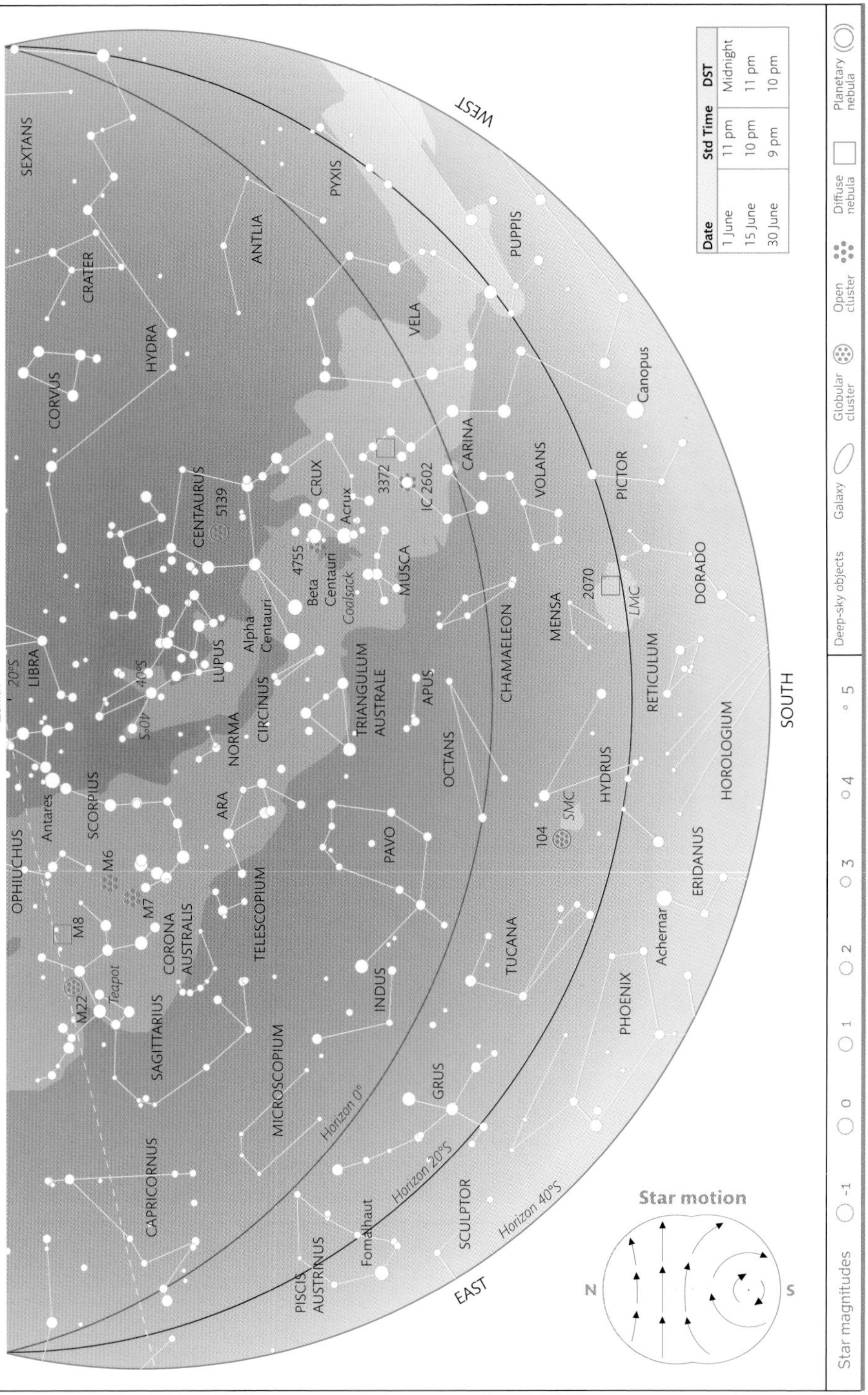

Date	Std Time	DST
1 June	11 pm	Midnight
15 June	10 pm	11 pm
30 June	9 pm	10 pm

WEST

SOUTH

EAST

Zenith

Horizon 0°
Horizon 20°S
Horizon 40°S

20°S
40°S
60°S

Constellations and objects

SEXTANS
CRATER
CORVUS
HYDRA
ANTLIA
PYXIS
PUPPIS
VELA
CARINA
Canopus
PICTOR
VOLANS
DORADO
3372
IC 2602
CRUX
Acrux
MUSCA
4755
Beta Centauri
Coalsack
Alpha Centauri
CENTAURUS
5139
LIBRA
LUPUS
NORMA
CIRCINUS
TRIANGULUM AUSTRALE
APUS
CHAMAELEON
MENSA
2070
LMC
RETICULUM
HOROLOGIUM
ERIDANUS
Achernar
PHOENIX
OCTANS
HYDRUS
SMC
104
TUCANA
PAVO
ARA
TELESCOPIUM
INDUS
GRUS
MICROSCOPIUM
SCULPTOR
PISCIS AUSTRINUS
Fomalhaut
CAPRICORNUS
SAGITTARIUS
M22
Teapot
CORONA AUSTRALIS
M7
M6
M8
SCORPIUS
Antares
OPHIUCHUS

Star motion

N   S

JULY

THIS IS THE BEST TIME for northern observers to
see the rich southern constellations Sagittarius and
Scorpius. The centre of our Galaxy is in Sagittarius,
so the Milky Way star fields are particularly dense in
this region. For southern observers, these constellations
are almost overhead and dominate the scene.

SUNRISE AND SUNSET ON 15 JULY

Latitude	Sunrise	Sunset
60°N	03.00	21.10
40°N	04.40	19.30
20°N	05.30	18.40
0°	06.00	18.10
20°S	06.40	17.40
40°S	07.20	16.50

1 July: Std time ⏱ DST ⏱	15 July: Std time ⏱ DST ⏱	30 July: Std time ⏱ DST ⏱

NORTHERN LATITUDES

LOOKING NORTH Ursa Minor (the Little Bear)
remains well placed, with Draco coiled around
it. The Plough (or Big Dipper) is in the
northwest, with Cassiopeia and Cepheus
in the northeast. The Great Square
of Pegasus is rising in the east, while
Arcturus, in Boötes, remains high
in the west.
LOOKING SOUTH Due south,
Ophiuchus is entwined by both
halves of Serpens with Hercules
above it. In the east, the Summer
Triangle of Vega, Deneb, and
Altair is well placed. Antares in
Scorpius is near the southern
horizon, with Sagittarius visible
to its left for observers in
southerly locations. Virgo
is setting in the far west.

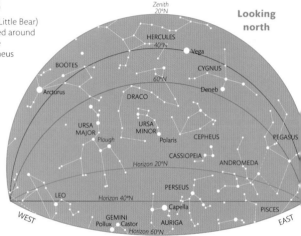

SOUTHERN LATITUDES

LOOKING NORTH Ophiuchus and both parts
of Serpens are centrally placed, with Aquila
to their right. Hercules is closer to the
northern horizon, midway between
reddish Arcturus in the northwest
and blue-white Vega in the north.
To the right of Vega, Cygnus is rising.
LOOKING SOUTH Scorpius,
shaped like a fish hook, and
Sagittarius are almost overhead,
containing dense fields of stars
near the centre of our Galaxy.
To the southwest, along the
star trail of the Milky Way (see
pp.184–85), are Centaurus and
Crux (the Southern Cross); Virgo
is setting further west. In the
east, Capricornus, Aquarius, and
the star Fomalhaut are rising.

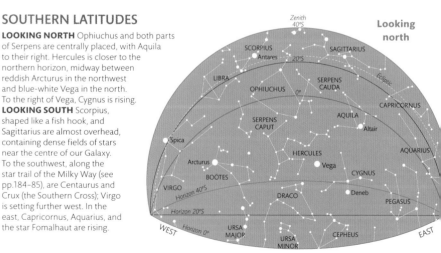

The Galactic Centre (Sagittarius)
From southern latitudes on July evenings, the centre of our Galaxy lies overhead, among dense Milky Way star fields in Sagittarius. Eight stars in Sagittarius (enclosed here in a rectangle) form a shape known as the Teapot (see pp.124–25).

WHOLE-SKY CHARTS ››

July features

• **Antares (Scorpius).** Also known as Alpha (α) Scorpii, this is one of the largest stars visible to the naked eye. It is a red supergiant, around 500 times larger than the Sun. If it were placed where our Sun is, it would engulf the orbit of Mars. See also pp.126–27.

• **Epsilon (ε) Lyrae.** Near the bright star Vega lies this remarkable multiple star. Binoculars, or even sharp eyesight, show that it is double. Through a telescope, each of these stars is itself seen to be a close double, hence this star's popular name, the Double Double. See also p.108.

☐ **M8 (Sagittarius).** This diffuse nebula is visible to the naked eye in country skies and is easy to find with binoculars. It is elongated in shape and is also called the Lagoon Nebula, the lagoon being a dark lane of dust that crosses its centre. M8 contains NGC 6530, a cluster visible through binoculars. See also pp.124–25.

⊛ **M22 (Sagittarius).** This globular cluster, which lies near the lid of the Teapot in Sagittarius, is rated the third best of its kind in the sky, after Omega (ω) Centauri and 47 Tucanae. It can be seen as a hazy star with the naked eye, and is easy to find with binoculars. See also pp.124–25.

Star magnitudes   ◯ -1   ◯ 0   ○ 1   ○ 2   · 3

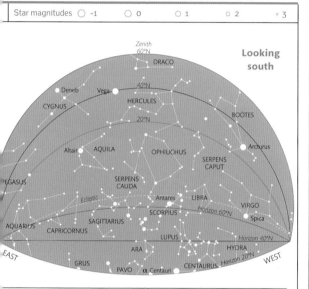

Looking south

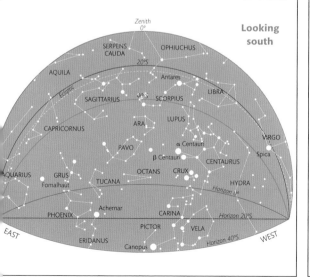

Looking south

Also visible
• **Albireo** (p.187)
⁛ **M6 and M7** (p.175)
⊛ **M13** (p.175)
⊛ **M27** (p.193)
• **The Cygnus Rift** (p.187)

WHOLE-SKY CHART

Northern latitudes

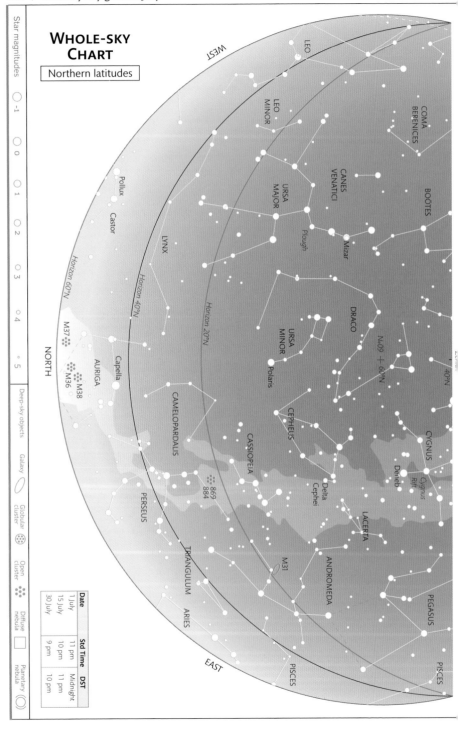

Star magnitudes

○ -1   ○ 0   ○ 1   ○ 2   ○ 3   ○ 4   ° 5

Deep-sky objects

Galaxy ⬭   Globular cluster 🌑   Open cluster ⁙   Diffuse nebula ▢   Planetary nebula ◎

NORTH

WEST

EAST

LEO

COMA BERENICES

BOOTES

LEO MINOR

CANES VENATICI

URSA MAJOR

Mizar

Plough

Pollux

Castor

LYNX

DRACO

N.09 + 60°N

CYGNUS

URSA MINOR

Polaris

Cygnus Rift

Deneb

M37

M38

M36

AURIGA

Capella

CAMELOPARDALIS

CEPHEUS

CASSIOPEIA

869 884

PERSEUS

Delta Cephei

LACERTA

ANDROMEDA

PEGASUS

TRIANGULUM

M31

PISCES

ARIES

PISCES

Horizon 60°N

Horizon 40°N

Horizon 20°N

40°N

Date	Std Time	DST
1 July	11 pm	Midnight
15 July	10 pm	11 pm
30 July	9 pm	10 pm

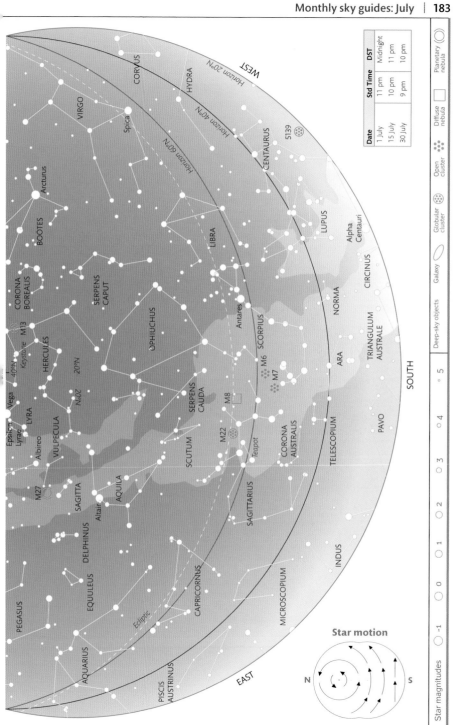

WEST

Horizon 20°N

Horizon 40°N

Horizon 60°N

CORVUS

HYDRA

VIRGO

Spica

Arcturus

BOOTES

CORONA
BOREALIS

M13
Keystone
40°N

Vega
Epsilon
Lyrae
LYRA

Albireo
VULPECULA

M27

SAGITTA

DELPHINUS

Altair
AQUILA

EQUULEUS

PEGASUS

AQUARIUS

PISCIS
AUSTRINUS

CAPRICORNUS

Ecliptic

MICROSCOPIUM

INDUS

SCUTUM

SERPENS
CAUDA

SAGITTARIUS

Teapot

M22

M8

OPHIUCHUS

SERPENS
CAPUT

HERCULES

20°N

Antares

M6
M7

SCORPIUS

LIBRA

CORONA
AUSTRALIS

TELESCOPIUM

PAVO

NORMA

ARA

LUPUS

Alpha
Centauri

CIRCINUS

TRIANGULUM
AUSTRALE

CENTAURUS

5139

EAST

SOUTH

Star motion

N   S

Date	Std Time	DST
1 July	11 pm	Midnight
15 July	10 pm	11 pm
30 July	9 pm	10 pm

Deep-sky objects

Galaxy | Globular cluster | Open cluster | Diffuse nebula | Planetary nebula

Star magnitudes

-1   0   1   2   3   4   5

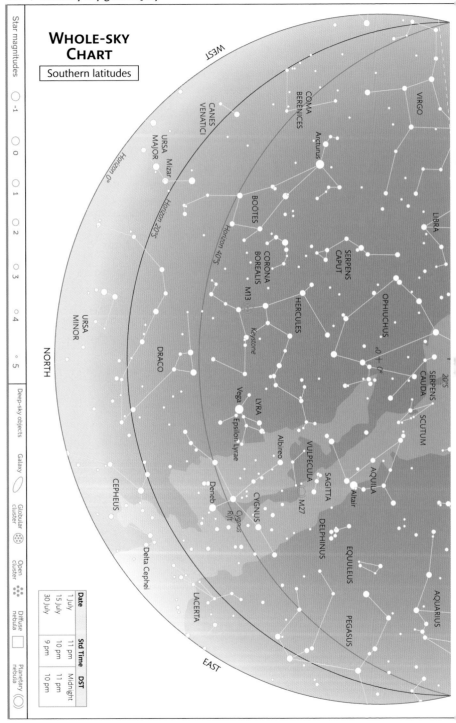

WHOLE-SKY CHART

Southern latitudes

Star magnitudes

○ -1
○ 0
○ 1
○ 2
○ 3
○ 4
○ 5

Deep-sky objects

Galaxy ◯
Globular cluster ⊛
Open cluster ⁘⁘
Diffuse nebula ☐
Planetary nebula ◎

WEST

NORTH

EAST

Horizon 0°
Horizon 20°S
Horizon 40°S

VIRGO
COMA BERENICES
Arcturus
CANES VENATICI
URSA MAJOR
Mizar
BOÖTES
LIBRA
SERPENS CAPUT
CORONA BOREALIS
M13
HERCULES
OPHIUCHUS
Keystone
URSA MINOR
+0° 0°
20°S
SERPENS CAUDA
SCUTUM
DRACO
Vega
LYRA
Epsilon Lyrae
Albireo
VULPECULA
AQUILA
Altair
SAGITTA
M27
DELPHINUS
CEPHEUS
Deneb
CYGNUS
Cygnus Rift
EQUULEUS
Delta Cephei
LACERTA
PEGASUS
AQUARIUS

Date	Std Time	DST
1 July	11 pm	Midnight
15 July	10 pm	11 pm
30 July	9 pm	10 pm

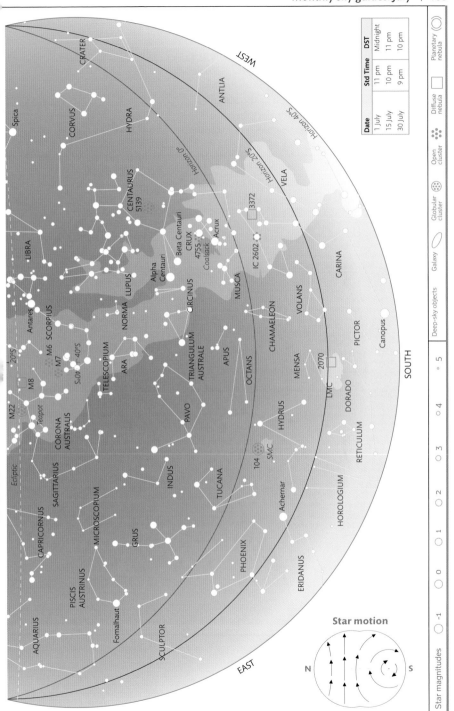

WEST

ANTLIA

CRATER

CORVUS

HYDRA

Spica

Horizon 0°S

LIBRA

CENTAURUS
5139

Horizon 20°S

VELA

Horizon 40°S

Beta Centauri
CRUX
4755
Coalsack
Acrux

Alpha
Centauri

CARINA

IC 2602

3372

Antares

SCORPIUS

NORMA
LUPUS

CIRCINUS

MUSCA

M6
M7

ARA

TELESCOPIUM

CHAMAELEON

VOLANS

30°S

M8
40°S

Ecliptic

CORONA
AUSTRALIS

Teapot

M22

SAGITTARIUS

CAPRICORNUS

MICROSCOPIUM

PAVO

TRIANGULUM
AUSTRALE

APUS

OCTANS

MENSA

2070

LMC

DORADO

PICTOR

Canopus

SOUTH

0 5

0 4

0 3

0 2

0 1

○ 0

○ -1

HYDRUS

SMC

104

INDUS

TUCANA

RETICULUM

HOROLOGIUM

Achernar

GRUS

PISCIS
AUSTRINUS

AQUARIUS

Fomalhaut

PHOENIX

ERIDANUS

SCULPTOR

EAST

Star motion

N

S

Star magnitudes

Deep-sky objects

Galaxy

Globular
cluster

Open
cluster

Diffuse
nebula

Planetary
nebula

Date	Std Time	DST
1 July	11 pm	Midnight
15 July	10 pm	11 pm
30 July	9 pm	10 pm

AUGUST

THE BRIGHT STARS Altair, Deneb, and Vega form a prominent triangle that is well seen on August evenings from all but the most southerly latitudes. In the northern hemisphere, it is known as the Summer Triangle. Another major sight for northern observers is the Perseid meteor shower, visible in mid-month.

SUNRISE AND SUNSET ON 15 AUGUST

Latitude	Sunrise	Sunset
60°N	04.10	19.50
40°N	05.10	17.00
20°N	05.40	18.30
0°	06.00	18.10
20°S	06.20	17.50
40°S	06.50	17.20

1 August: Std time ⏱ DST ⏱	15 August: Std time ⏱ DST ⏱	30 August: Std time ⏱ DST ⏱

NORTHERN LATITUDES

LOOKING NORTH The Plough (or Big Dipper) is in the northwest with Arcturus to its left. The bowl of the Little Dipper (in Ursa Minor) is to the left of Polaris, while to the right Cassiopeia appears like a back-to-front letter E. Perseus is rising lower down, with Pegasus and Andromeda further east.
LOOKING SOUTH The stars of the Summer Triangle – Vega, Deneb, and Altair – are overhead. The Milky Way (see pp.188–89) runs southwest to northeast, and seems to divide in two in Cygnus. Ophiuchus and both parts of Serpens are in the west, with Sagittarius and Scorpius beneath them. Aquarius and Capricornus are in the southeast.

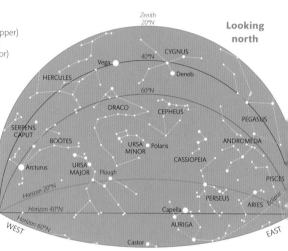

SOUTHERN LATITUDES

LOOKING NORTH The bright stars Vega, Altair (in Aquila), and Deneb (in Cygnus) form a large triangle due north. The Milky Way (see pp.190–91), running northeastwards from near the zenith, seems to split in Cygnus. Ophiuchus and both parts of Serpens dominate the western sky, with Hercules just beneath them. In the northeast, the Great Square of Pegasus is rising.
LOOKING SOUTH In the southeast, Achernar is rising, with Fomalhaut above it and slightly to the left. Higher up are Capricornus and Aquarius. Sagittarius is near the zenith, with Scorpius to its lower right. Centaurus and Crux (the Southern Cross) are setting in the southwest.

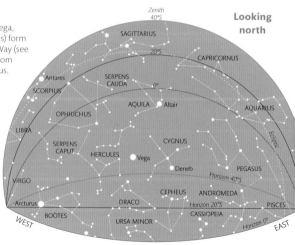

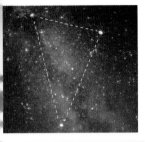

The Summer Triangle
Three stars in three separate constellations form a large isosceles triangle, which is a familiar feature in skies of northern summer and autumn, and southern winter and spring. The brightest of the three is Vega (top right) in Lyra, followed by Altair (bottom) in Aquila, and Deneb (top left) in Cygnus. The Cygnus Rift (see panel, right) runs through the triangle.

Star magnitudes ◯ -1 ◯ 0 ◯ 1 ○ 2 ∘ 3

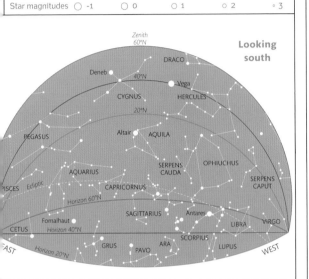

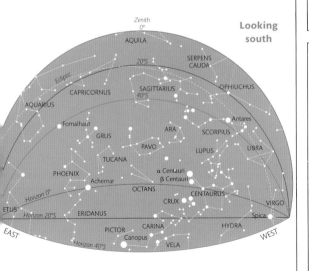

WHOLE-SKY CHARTS »

August features

● **Albireo (Cygnus).** This star, also known as Beta (β) Cygni, is one of the most celebrated doubles in the sky, on account of the coloration of its component stars and the ease with which they may be separated. The smallest of telescopes will show the stars; they appear amber and greenish, like two lamps in a celestial traffic light. See also p.88.

● **The Cygnus Rift (Cygnus).** This dark nebula, also known as the Northern Coalsack, divides the Milky Way into two. It can be traced with the naked eye from Cygnus, through Aquila, and into Ophiuchus, where it broadens out. See also p.88.

Also visible
● **Antares** (p.181)
● **Delta (δ) Cephei** (p.193)
● **Epsilon (ε) Lyrae** (p.181)
☐ **M8** (p.181)
✸ **M13** (p.175)
✸ **M22** (p.181)
◯ **M27** (p.193)
✸ **NGC 104** (p.193)

August meteors

The Perseids. Observers in the northern hemisphere have an opportunity this month to see the finest meteor shower of the year, the Perseids, which radiate from near the Double Cluster (NGC 869 and 884) in Perseus (see p.118). The meteors are bright, can flare up, and often leave trains. A peak of 75 or more meteors an hour is reached around 12 August, and activity can be seen for at least a week either side of this date.

WHOLE-SKY CHART

Northern latitudes

WEST

COMA BERENICES

Arcturus

LEO MINOR

CANES VENATICI

BOÖTES

CORONA BOREALIS

HERCULES

M13

URSA MAJOR

Mizar

Vega

Epsilon Lyrae

Plough

DRACO

CYGNUS

N69 + 60°N

Cygnus Rift

Deneb

Horizon 60°N

LYNX

Horizon 40°N

URSA MINOR

Horizon 20°N

LACERTA

Polaris

CEPHEUS

PEGASUS

Castor

CAMELOPARDALIS

CASSIOPEIA

Delta Cephei

NORTH

869884

M31

Capella

ANDROMEDA

AURIGA

M38
M36

TRIANGULUM

M37

PERSEUS

ARIES

PISCES

TAURUS

Pleiades

EAST

Star magnitudes

○ -1
○ 0
○ 1
○ 2
○ 3
○ 4
° 5

Deep-sky objects

Galaxy ⬭

Globular cluster ⊛

Open cluster ⣿

Diffuse nebula ▢

Planetary nebula ◯

Date	Std Time	DST
1 August	11 pm	Midnight
15 August	10 pm	11 pm
30 August	9 pm	10 pm

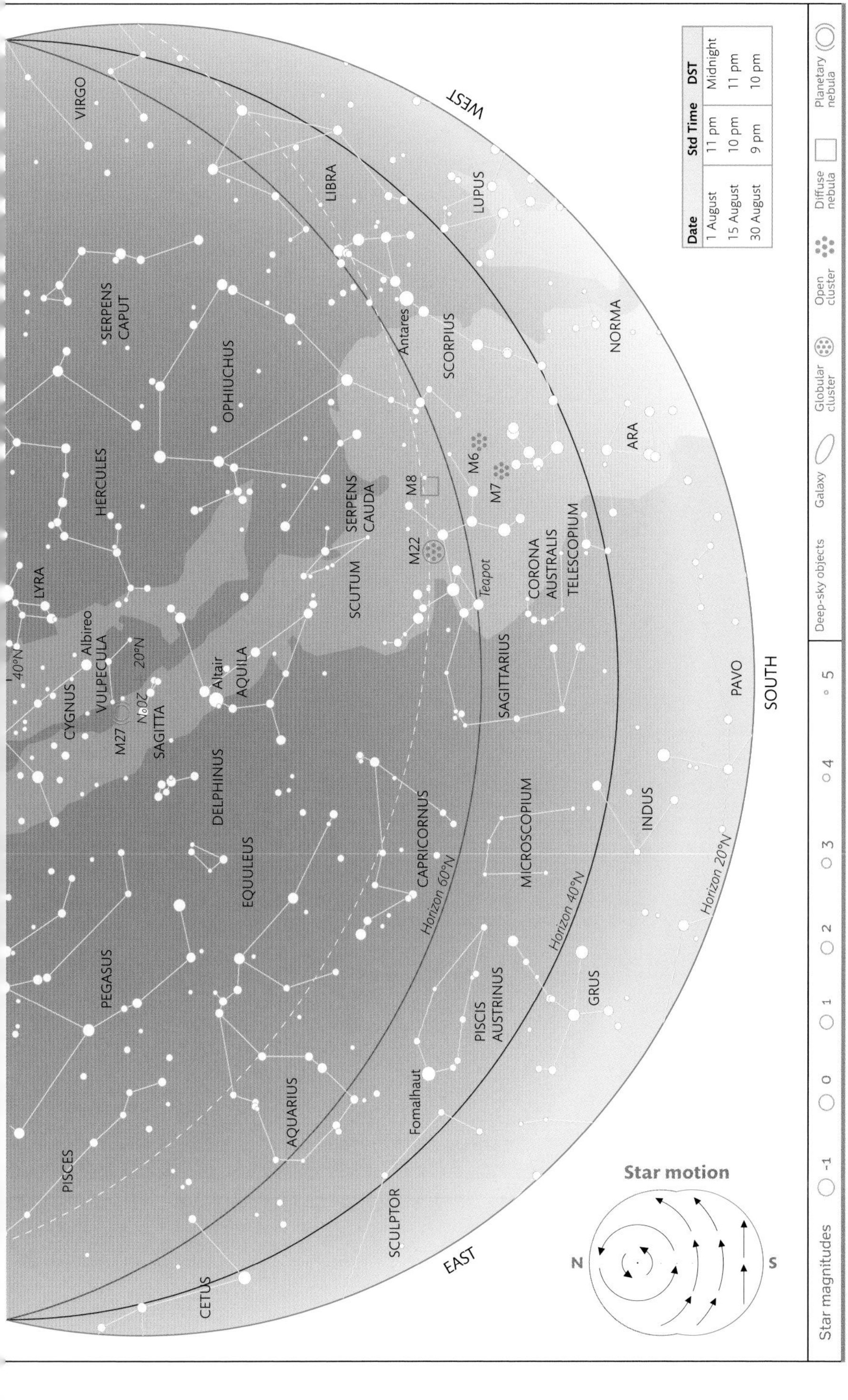

Date	Std Time	DST
1 August	11 pm	Midnight
15 August	10 pm	11 pm
30 August	9 pm	10 pm

Star motion

Star magnitudes

○ -1   ○ 0   ○ 1   ○ 2   ○ 3   ○ 4   ○ 5

Deep-sky objects   Galaxy   Globular cluster   Open cluster   Diffuse nebula   Planetary nebula

WHOLE-SKY CHART

Southern latitudes

Star magnitudes

○ -1
○ 0
○ 1
○ 2
○ 3
○ 4
° 5

Deep-sky objects

Galaxy ⬭
Globular cluster ⊛
Open cluster ⁙
Diffuse nebula ▢
Planetary nebula ◎

WEST

NORTH

EAST

Horizon 0°
Horizon 20°S
Horizon 40°S

Arcturus

BOOTES

CORONA BOREALIS

SERPENS CAPUT

OPHIUCHUS

SERPENS CAUDA

HERCULES

M13

SCUTUM

DRACO

Vega
LYRA
Epsilon Lyrae

VULPECULA
Albireo
M27

AQUILA
Altair

AQUARIUS

URSA MINOR

SAGITTA

DELPHINUS

EQUULEUS

Deneb

CYGNUS
Cygnus Rift

CEPHEUS

LACERTA

Delta Cephei

PEGASUS

CASSIOPEIA

ANDROMEDA

M31

PISCES

Ecliptic

20°S

0°

Date	Std Time	DST
1 August	11 pm	Midnight
15 August	10 pm	11 pm
30 August	9 pm	10 pm

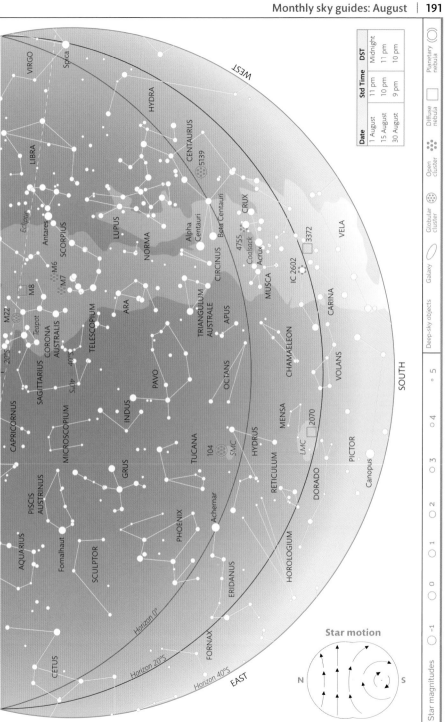

Date	Std Time	DST
1 August	11 pm	Midnight
15 August	10 pm	11 pm
30 August	9 pm	10 pm

Deep-sky objects

Globular cluster

Open cluster

Galaxy

Diffuse nebula

Planetary nebula

Star magnitudes

-1 0 1 2 3 4 5°

WEST

SOUTH

EAST

Star motion

N   S

VIRGO
Spica
HYDRA
LIBRA
CENTAURUS
5139
CRUX
Ecliptic
Antares
SCORPIUS
Alpha Centauri
Beta Centauri
4755
Coalsack
Acrux
LUPUS
NORMA
CIRCINUS
MUSCA
3372
VELA
M6
M7
M8
ARA
TRIANGULUM AUSTRALE
APUS
CARINA
IC 2602
M22
Teapot
CORONA AUSTRALIS
TELESCOPIUM
PAVO
OCTANS
CHAMAELEON
20°S
SAGITTARIUS
40°S
30°S
VOLANS
CAPRICORNUS
MICROSCOPIUM
INDUS
MENSA
2070
PICTOR
GRUS
TUCANA
104
SMC
HYDRUS
LMC
PISCIS AUSTRINUS
RETICULUM
DORADO
Fomalhaut
PHOENIX
Achernar
Canopus
AQUARIUS
SCULPTOR
HOROLOGIUM
CETUS
ERIDANUS
FORNAX
Horizon 0°
Horizon 20°S
Horizon 40°S

SEPTEMBER

DAYS AND NIGHTS BECOME EQUAL in length as the equinox approaches, around 23 September, marking the start of northern autumn and southern spring. The Great Square of Pegasus is well placed for observers in northern latitudes, while in the southern hemisphere the brightest stars are in the western sky.

SUNRISE AND SUNSET ON 15 SEPTEMBER

Latitude	Sunrise	Sunset
60°N	05.30	18.20
40°N	05.40	18.10
20°N	05.50	18.00
0°	05.50	18.00
20°S	06.00	18.00
40°S	06.00	17.50

1 September: Std time ☉ DST ☉	15 September: Std time ☉ DST ☉	30 September: Std time ☉ DST ☉

NORTHERN LATITUDES

• **LOOKING NORTH** The view to the north contains relatively few bright stars this month. Cepheus is due north, with Cassiopeia to its right. Andromeda and Pegasus are high in the eastern sky, while Perseus is rising in the northeast, followed by Auriga. Draco and Hercules are in the northwestern sky.

• **LOOKING SOUTH** Aquarius and Capricornus are centrally placed, with Fomalhaut below them. The Great Square of Pegasus, with Pisces to its lower left, dominates the view to the southeast. The Summer Triangle of Altair, Deneb, and Vega is almost overhead in the southwest, while Ophiuchus and Serpens Cauda set further west.

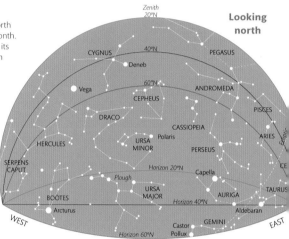

SOUTHERN LATITUDES

• **LOOKING NORTH** The large triangle formed by the bright stars Altair, Deneb, and Vega is prominent in the northwest. The Great Square of Pegasus dominates the northeastern sky, with Andromeda closer to the horizon and Pisces further east. Capricornus and Aquarius are almost overhead. Ophiuchus and Serpens Cauda are setting in the west.

• **LOOKING SOUTH** Achernar in Eridanus is prominent in the southeast, an otherwise fairly blank area of sky, with the Small Magellanic Cloud (SMC, see p.197) to its right. Fomalhaut is almost overhead. Scorpius is in the southwest, with Sagittarius above it.

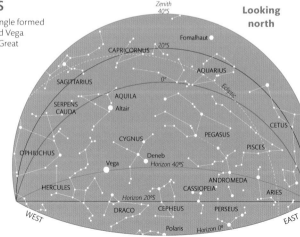

The Small Magellanic Cloud (Tucana)
This is the smaller, fainter, and more distant of two irregularly shaped small galaxies accompanying our own. To the naked eye, it looks like a cut-off part of the Milky Way. The cluster to its right is NGC 104 (see panel, right).

Star magnitudes ◯ -1   ◯ 0   ◯ 1   ◦ 2   ° 3

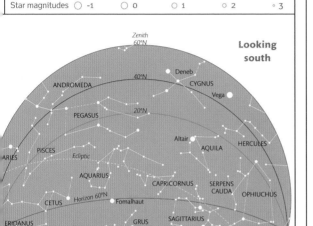

Looking south

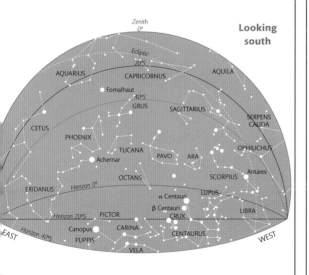

Looking south

WHOLE-SKY CHARTS »

September features

● **Delta (δ) Cephei.** This is the prototype of the Cepheid variable stars that are used for finding distances in space. Delta Cephei's brightness rises and falls every 5 days as the star pulsates in size; at its peak, it is over twice as bright as when at its faintest. Changes in its brightness can be followed with the naked eye by comparing it with nearby stars of known magnitude. Delta Cephei is also an attractive double star, having a fainter companion that is visible with the smallest telescopes. See also p.80.

(◯) **M27 (Vulpecula).** The easiest planetary nebula to see with binoculars lies in the often-overlooked constellation of Vulpecula, midway between the bright stars Deneb and Altair. Its popular name, the Dumbbell Nebula, comes from its structure – more like a bow tie than a dumbbell – but this is visible only through a telescope. See also p.141.

⊛ **NGC 104 (Tucana).** Also known as 47 Tucanae, this is the second most prominent globular cluster in the entire sky – only Omega (ω) Centauri is brighter. To the naked eye, it resembles a hazy star. It appears in the sky near the Small Magellanic Cloud (SMC) but is actually within our own Galaxy. See also p.135.

Also visible
● **Albireo** (p.187)
● **Epsilon (ε) Lyrae** (p.181)
◌ **M31** (p.199)
⁙ **NGC 869 and NGC 884** (p.199)
● **The Cygnus Rift** (p.187)

WHOLE-SKY CHART

Northern latitudes

Star magnitudes

○ -1
○ 0
○ 1
○ 2
○ 3
○ 4
○ 5

Deep-sky objects

⬭ Galaxy

⊛ Globular cluster

⣿ Open cluster

▫ Diffuse nebula

◎ Planetary nebula

NORTH

WEST

EAST

Arcturus

COMA BERENICES

SERPENS CAPUT

BOOTES

CORONA BOREALIS

HERCULES

CANES VENATICI

M13

Mizar

LYRA

Vega
Epsilon Lyrae

LEO MINOR

URSA MAJOR

DRACO

CYGNUS

Deneb

Cygnus Rift

*40°N

LACERTA

Plough

URSA MINOR

CEPHEUS

N 09° + 60°N Delta Cephei.

Polaris

Horizon 40°S

CASSIOPEIA

ANDROMEDA

LYNX

Horizon 20°S

CAMELOPARDALIS

869
884

M31

Horizon 0°

Castor

Pollux

Capella

AURIGA

PERSEUS

TRIANGULUM

GEMINI

M38 M36

M37

Pleiades

ARIES

TAURUS

Hyades

Ecliptic

Aldebaran

Date	Std Time	DST
1 September	11 pm	Midnight
15 September	10 pm	11 pm
30 September	9 pm	10 pm

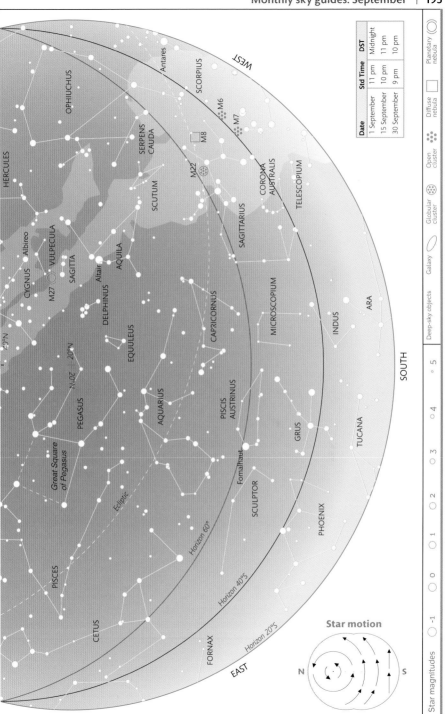

Date	Std Time	DST
1 September	11 pm	Midnight
15 September	10 pm	11 pm
30 September	9 pm	10 pm

WEST

SOUTH

EAST

Star motion

N   S

Star magnitudes   ○ -1   ○ 0   ○ 1   ○ 2   ○ 3   ○ 4   ○ 5

Deep-sky objects   Galaxy   Globular cluster   Open cluster   Diffuse nebula   Planetary nebula

HERCULES
OPHIUCHUS
SERPENS CAUDA
SCUTUM
SAGITTARIUS
CORONA AUSTRALIS
TELESCOPIUM
Antares
SCORPIUS
M6
M7
M8
M22
CYGNUS
Albireo
VULPECULA
M27
SAGITTA
AQUILA
Altair
DELPHINUS
EQUULEUS
PEGASUS
Great Square of Pegasus
Ecliptic
40°N
20°N
20°N
CAPRICORNUS
AQUARIUS
PISCIS AUSTRINUS
MICROSCOPIUM
INDUS
ARA
Fomalhaut
GRUS
TUCANA
SCULPTOR
PHOENIX
PISCES
CETUS
FORNAX
Horizon 60°
Horizon 40°S
Horizon 20°S

WHOLE-SKY CHART

Southern latitudes

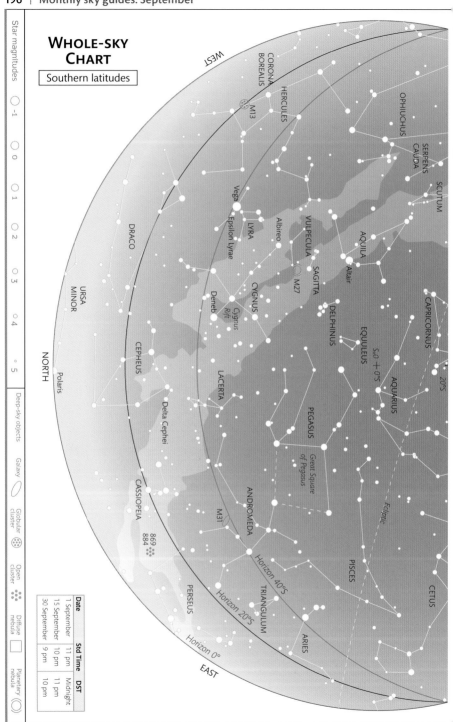

Star magnitudes

○	-1
○	0
○	1
○	2
○	3
○	4
∘	5

Deep-sky objects

Galaxy	◯
Globular cluster	⊛
Open cluster	⁂
Diffuse nebula	▢
Planetary nebula	◉

NORTH

WEST

EAST

Horizon 0°
Horizon 20°S
Horizon 40°S

Ecliptic

CORONA BOREALIS
HERCULES
M13
OPHIUCHUS
SERPENS CAUDA
SCUTUM
Vega
LYRA
Epsilon Lyrae
Albireo
VULPECULA
SAGITTA
M27
AQUILA
Altair
CAPRICORNUS
20°S
DRACO
CYGNUS
Cygnus Rift
Deneb
DELPHINUS
EQUULEUS
AQUARIUS
50°S
URSA MINOR
Polaris
CEPHEUS
LACERTA
PEGASUS
Great Square of Pegasus
Delta Cephei
CASSIOPEIA
ANDROMEDA
PISCES
CETUS
869 884
M31
TRIANGULUM
ARIES
PERSEUS

Date	Std Time	DST
1 September	11 pm	Midnight
15 September	10 pm	11 pm
30 September	9 pm	10 pm

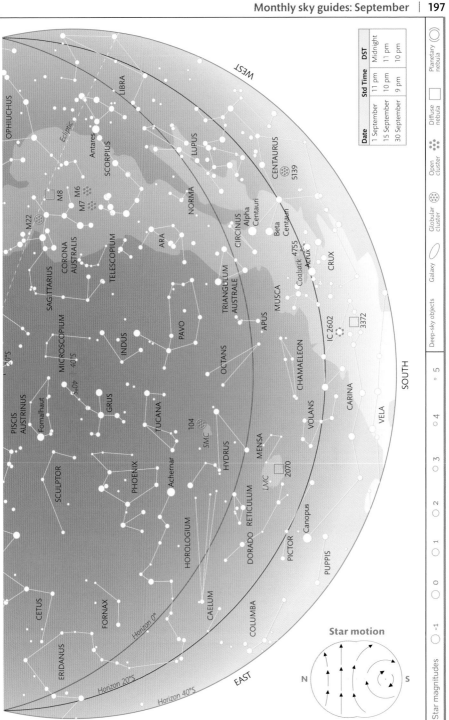

Date	Std Time	DST
1 September	11 pm	Midnight
15 September	10 pm	11 pm
30 September	9 pm	10 pm

WEST

OPHIUCHUS

LIBRA

Ecliptic

Antares

SCORPIUS

LUPUS

CENTAURUS
5139

M8
M6
M7

M22

NORMA

CORONA
AUSTRALIS

CIRCINUS

Alpha
Centauri

Beta
Centauri

Coalsack 4755
Acrux
CRUX

SAGITTARIUS

TELESCOPIUM

ARA

TRIANGULUM
AUSTRALE

MUSCA

30°S

MICROSCOPIUM

40°S

PAVO

INDUS

APUS

OCTANS

CHAMAELEON

IC 2602
3372

PISCIS
AUSTRINUS

Fomalhaut

GRUS

TUCANA

CARINA

SOUTH

VOLANS

VELA

SCULPTOR

PHOENIX

104

SMC

HYDRUS

MENSA

Achernar

RETICULUM

LMC
2070

DORADO

Canopus

PICTOR

HOROLOGIUM

CETUS

FORNAX

CAELUM

COLUMBA

PUPPIS

ERIDANUS

Horizon 0°S

Horizon 20°S

EAST

Horizon 40°S

Star motion

N   S

Deep-sky objects

Galaxy Globular Open Diffuse Planetary
        cluster   cluster nebula   nebula

° 5   ○ 4   ○ 3   ○ 2   ○ 1   ○ 0   ○ -1

Star magnitudes

OCTOBER

THE ANDROMEDA GALAXY, M31, is almost overhead for observers in northern latitudes, although for those in the southern hemisphere it remains disappointingly low. There are few bright stars on display at this time of year, but the Great Square of Pegasus is easy to see from both hemispheres.

SUNRISE AND SUNSET ON 15 OCTOBER

Latitude	Sunrise	Sunset
60°N	06.40	16.50
40°N	06.10	17.20
20°N	05.50	17.40
0°	05.40	17.50
20°S	05.30	18.00
40°S	05.10	18.20

1 October: Std time ⏱ DST ⏱	15 October: Std time ⏱ DST ⏱	30 October: Std time ⏱ DST ⏱

NORTHERN LATITUDES

LOOKING NORTH The Milky Way arches overhead from west to east (see pp.200–201). Cassiopeia is high up, above and to the left of Perseus. Capella and the other stars of Auriga are rising in the northeast along with Taurus, a foretaste of winter. The Summer Triangle can still be seen in the west, with Altair lying in the southwestern sky (far right of looking south map).

LOOKING SOUTH High in the south, the Great Square of Pegasus is well placed, with Andromeda above and to its left, almost overhead. Pisces lies to the left of Pegasus, and Aquarius is to its lower right. Low in the south is Fomalhaut, with Cetus in the southeastern sky.

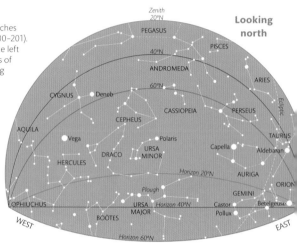

SOUTHERN LATITUDES

LOOKING NORTH The Great Square of Pegasus is centrally placed, with Andromeda to its lower right. Pisces is above and to the right of Pegasus, with Aquarius above left, nearly overhead. Altair, Vega, and Deneb are setting in the northwest. Aldebaran and the other stars of Taurus are just starting to appear above the northeastern horizon.

LOOKING SOUTH Achernar is high in the southeast with Canopus rising below it. The only other objects of note in this relatively blank area are the Large and Small Magellanic Clouds (LMC and SMC, see p.203). Low in the southwestern sky, Sagittarius is following Scorpius below the horizon.

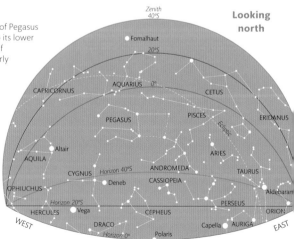

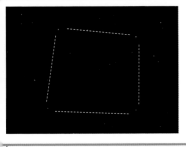

The Great Square of Pegasus
The corners of this huge square are marked by three stars in Pegasus and one in Andromeda (top left). Despite the figure's considerable size (about 15 degrees to a side, large enough to enclose 900 full Moons), it contains few naked-eye stars.

Star magnitudes	◯ -1	◯ 0	◯ 1	○ 2	° 3

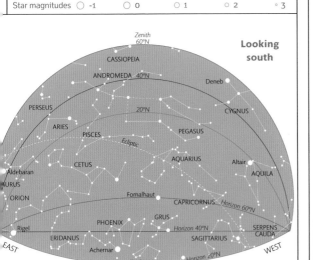

Looking south

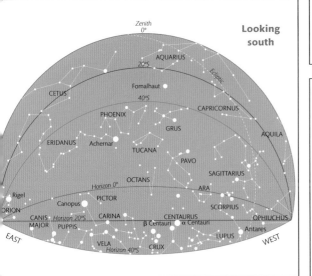

Looking south

WHOLE-SKY CHARTS ››

October features

🌀 **M31 (Andromeda).** The most distant object visible to the naked eye is a spiral galaxy similar to our home galaxy. Also called the Andromeda Galaxy, it can be seen with the naked eye in rural skies, but its full extent is better appreciated through binoculars. M31 is about 2.5 million light years away, so the light by which we now see it left when mankind's ancestors still roamed the plains of Africa. See also p.64.

⋰ **NGC 869 and NGC 884 (Perseus).** Two open clusters, known as the Double Cluster, mark the hand of Perseus. When seen with the naked eye, the pair resembles a knot that appears brighter than the surrounding Milky Way. They are easy to see through binoculars. See also p.118.

Also visible

- **Albireo** (p.187)
- **Delta (δ) Cephei** (p.193)
- ◯ **M27** (p.193)
- ⊕ **NGC 104** (p.193)
- ☐ **NGC 2070** (p.205)
- **The Cygnus Rift** (p.187)
- **The Hyades and Pleiades** (p.211)

October meteors

The Orionids. Radiating from northern Orion, near the border with Gemini, the Orionids reach a peak of about 25 meteors an hour around 21 October. As this area of sky does not rise until late, the shower is best viewed after midnight. The Orionids are fast moving but faint. Like the Eta Aquariids of May, they are caused by dust from Halley's Comet.

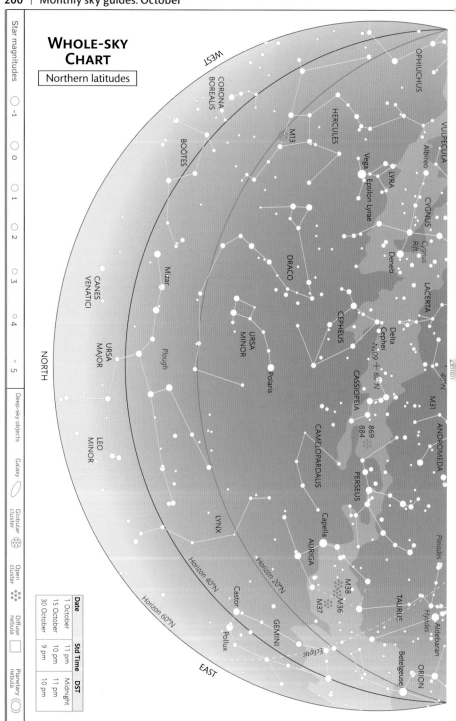

WHOLE-SKY CHART

Northern latitudes

Date	Std Time	DST
1 October	11 pm	Midnight
15 October	10 pm	11 pm
30 October	9 pm	10 pm

Star magnitudes: -1, 0, 1, 2, 3, 4, 5

Deep-sky objects — Galaxy, Globular cluster, Open cluster, Diffuse nebula, Planetary nebula

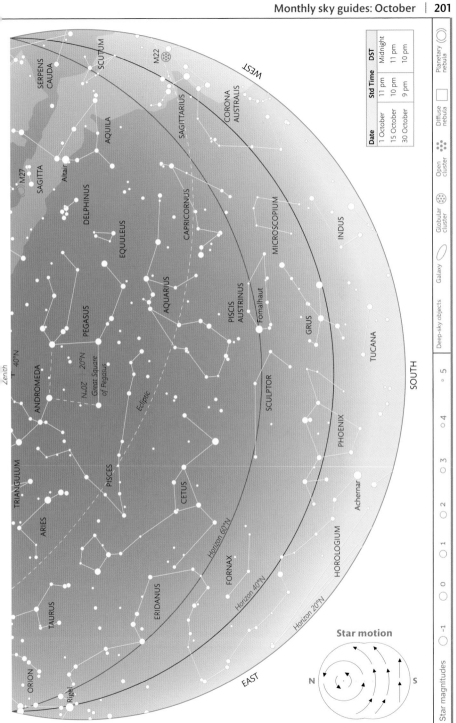

Date	Std Time	DST
1 October	11 pm	Midnight
15 October	10 pm	11 pm
30 October	9 pm	10 pm

Deep-sky objects

Galaxy · Globular cluster · Open cluster · Diffuse nebula · Planetary nebula

Star magnitudes

-1 · 0 · 1 · 2 · 3 · 4 · 5 °

Star motion

WEST

SOUTH

EAST

SERPENS CAUDA
SCUTUM
M22
SAGITTA
AQUILA
Altair
M27
SAGITTARIUS
CORONA AUSTRALIS
DELPHINUS
CAPRICORNUS
EQUULEUS
MICROSCOPIUM
INDUS
PEGASUS
AQUARIUS
PISCIS AUSTRINUS
Fomalhaut
GRUS
Great Square of Pegasus
20°N
40°N
Zenith
Ecliptic
ANDROMEDA
TUCANA
TRIANGULUM
PISCES
SCULPTOR
PHOENIX
ARIES
CETUS
Achernar
TAURUS
FORNAX
Horizon 60°N
HOROLOGIUM
ORION
ERIDANUS
Horizon 40°N
Rigel
Horizon 20°N

WHOLE-SKY CHART

Southern latitudes

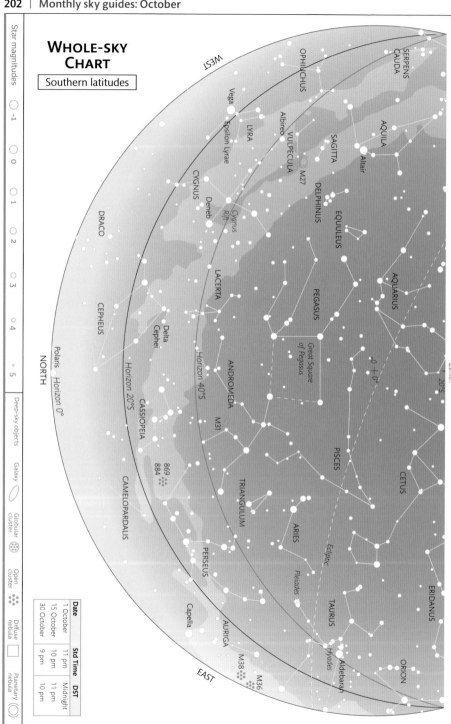

Star magnitudes

○ -1
○ 0
○ 1
○ 2
○ 3
○ 4
° 5

Deep-sky objects

⬭ Galaxy
⊛ Globular cluster
⁙ Open cluster
▫ Diffuse nebula
◎ Planetary nebula

Date	Std Time	DST
1 October	11 pm	Midnight
15 October	10 pm	11 pm
30 October	9 pm	10 pm

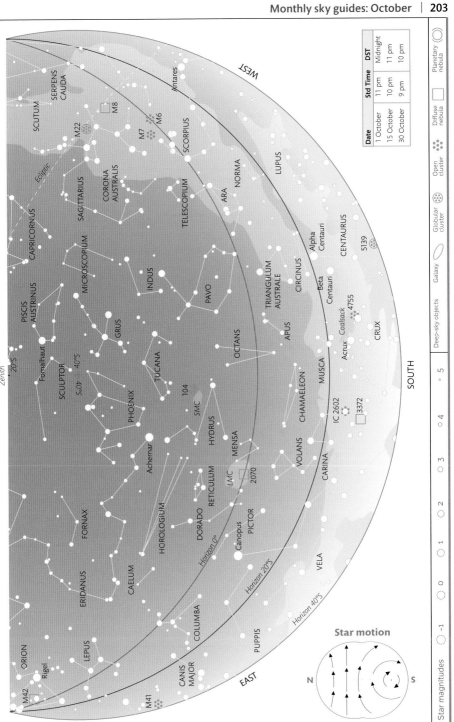

Star motion

Deep-sky objects

Date	Std Time	DST
1 October	11 pm	Midnight
15 October	10 pm	11 pm
30 October	9 pm	10 pm

Star magnitudes ● -1 ● 0 ○ 1 ○ 2 ○ 3 ○ 4 ○ 5

Galaxy · Globular cluster · Open cluster · Diffuse nebula · Planetary nebula

NOVEMBER

ALL THE constellations named after characters in the story of Perseus and Andromeda are on view this month – Perseus and Andromeda themselves, Cassiopeia and Cepheus, her parents, and Cetus, the sea monster from which she was saved by Perseus. In northern latitudes, daylight-saving time is at an end.

SUNRISE AND SUNSET ON 15 NOVEMBER

Latitude	Sunrise	Sunset
60°N	08.00	15.30
40°N	06.50	16.40
20°N	06.10	17.20
0°	05.40	17.50
20°S	05.10	18.20
40°S	04.40	18.50

1 November: Std time ◷ DST ◷	15 November: Std time ◷ DST ◷	30 November: Std time ◷ DST ◷

NORTHERN LATITUDES

LOOKING NORTH The Milky Way (see pp.206–207) arches overhead from west to east. Perseus, Andromeda, and Cassiopeia are high up, with Cepheus beneath them. Auriga, with its brightest star Capella, is high in the northeast, with Gemini below it. The Summer Triangle of Vega, Deneb, and Altair is now low in the northwestern sky.

LOOKING SOUTH Cetus lies due south, flanked higher up by Aries and Andromeda. Pisces, Pegasus, and Aquarius dominate the sky to the west. The stars of winter are rising in the east, among them Orion and Taurus, which contains red Aldebaran and the Hyades and Pleiades clusters (see p.207).

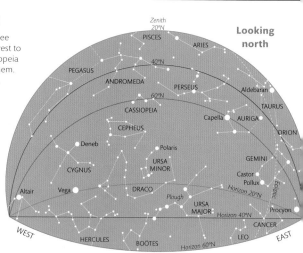

SOUTHERN LATITUDES

LOOKING NORTH Cetus is almost overhead, with Pisces and Aries below it, while Perseus, Andromeda, and Cassiopeia are close to the northern horizon. Pegasus, with its Great Square, occupies most of the northwestern sky. In the northeast, Taurus is climbing into view, with Orion further east, a sign of approaching summer.

LOOKING SOUTH Achernar is due south, with the Large and Small Magellanic Clouds (LMC and SMC, see p.209) below it. There are few noticeable stars near the zenith, but Sirius and Canopus, the two brightest stars in the sky, stand out in the east and southeast. Fomalhaut is high in the west.

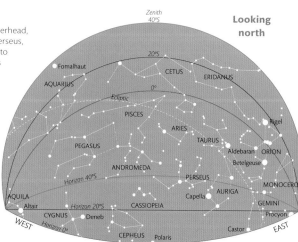

The Large Magellanic Cloud

Lying mostly in Dorado, this is the larger, brighter, and closer of two small galaxies that accompany our own. It appears to the eye as a long, hazy patch like a detached part of the Milky Way, while binoculars show a host of star clusters and nebulae.

WHOLE-SKY CHARTS »

November features

☐ **NGC 2070 (Dorado).**
The most prominent object in the Large Magellanic Cloud is a diffuse nebula visible to the naked eye and through binoculars. It is popularly known as the Tarantula Nebula because of its spidery shape, which is particularly noticeable on photographs. It is about 50 times larger than the famous Orion Nebula (M42) in our own Galaxy. See also p.91.

Also visible

🌑 **M31** (p.199)
⁂ **M36, M37, and M38** (p.211)
⁂ **M41** (p.151)
☐ **M42** (p.145)
⁂ **NGC 869 and NGC 884** (p.199)
⁂ **NGC 2244** (p.151)
● **The Hyades and Pleiades** (p.211)

Star magnitudes	○ -1	○ 0	○ 1	○ 2	∘ 3

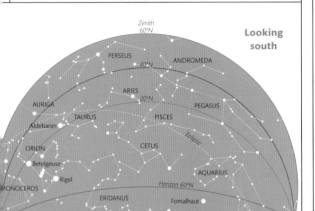

Looking south

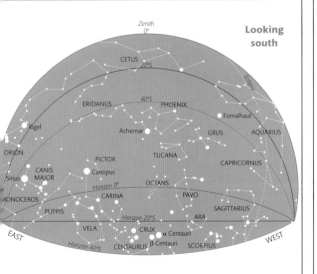

Looking south

November meteors

The Taurids. Starting in late October and continuing for a month or more, the Taurids reach a sustained peak in the first week of November. Activity is no more than 10 an hour, from just south of the Pleiades, but Taurid meteors are striking because they move slowly and are often bright.

The Leonids. This shower radiates from the head of Leo around 17 November. Activity is normally only about 10 an hour at best, but it can surge sharply at 33-year intervals when the parent comet, Tempel-Tuttle, returns to the Sun. A Leonid storm occurred in 1966, and there were high rates of activity from 1998 to 2002.

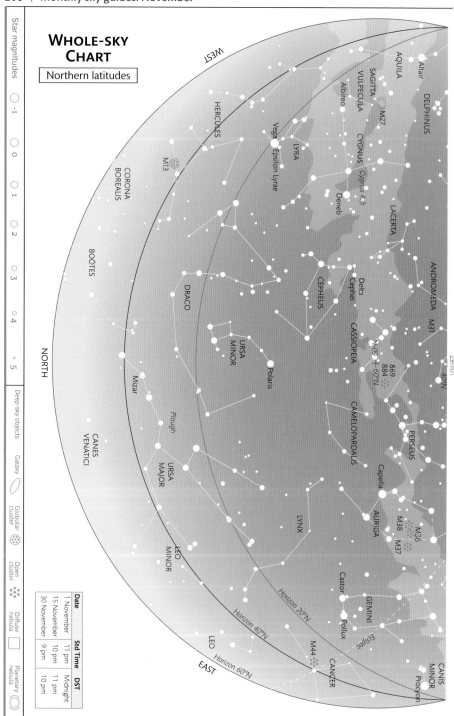

WHOLE-SKY CHART

Northern latitudes

Star magnitudes

○ -1
○ 0
○ 1
○ 2
○ 3
○ 4
° 5

NORTH

Deep-sky objects

Galaxy ⬭

Globular cluster ⊛

Open cluster ⸭

Diffuse nebula ☐

Planetary nebula ◎

Date	Std Time	DST
1 November	11 pm	Midnight
15 November	10 pm	11 pm
30 November	9 pm	10 pm

WEST

AQUILA
Altair
DELPHINUS
SAGITTA
VULPECULA
M27
Albireo
CYGNUS
Cygnus Rift
HERCULES
Vega
LYRA
Epsilon Lyrae
Deneb
LACERTA
M13
CORONA BOREALIS
ANDROMEDA
M31
Delta Cephei
CEPHEUS
CASSIOPEIA
N.09
60°N
869
884
zenith
40°N
BOÖTES
DRACO
URSA MINOR
Polaris
CAMELOPARDALIS
PERSEUS
Mizar
Plough
URSA MAJOR
Capella
CANES VENATICI
AURIGA
M38
M36
M37
LYNX
Castor
GEMINI
LEO MINOR
Pollux
Ecliptic
M44
CANCER
CANIS MINOR
Procyon
LEO
Horizon 20°N
Horizon 40°N
Horizon 60°N
EAST

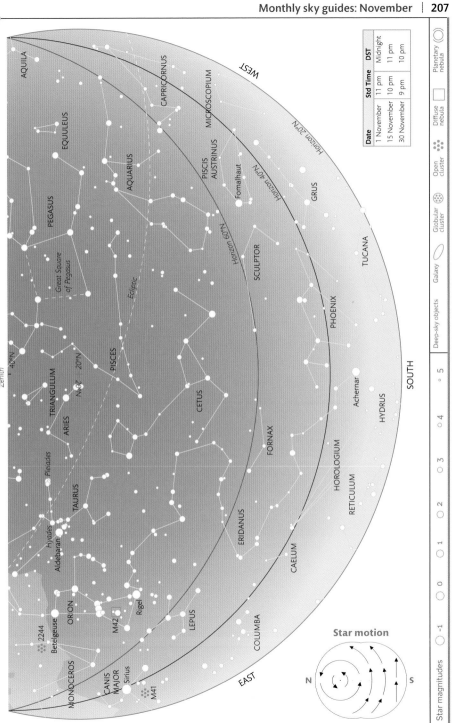

Date	Std Time	DST
1 November	11 pm	Midnight
15 November	10 pm	11 pm
30 November	9 pm	10 pm

Deep-sky objects

Galaxy · Globular cluster · Open cluster · Diffuse nebula · Planetary nebula

Star magnitudes: -1, 0, 1, 2, 3, 4, 5

Star motion

N S

AQUILA
EQUULEUS
PEGASUS
Great Square of Pegasus
Ecliptic
TRIANGULUM
ARIES
Pleiades
TAURUS
Hyades
Aldebaran
2244
Betelgeuse
ORION
M42
Rigel
MONOCEROS
CANIS MAJOR
Sirius
M41
LEPUS
COLUMBA
CAPRICORNUS
MICROSCOPIUM
PISCIS AUSTRINUS
AQUARIUS
Fomalhaut
GRUS
TUCANA
SCULPTOR
PISCES
CETUS
PHOENIX
Achernar
HYDRUS
FORNAX
HOROLOGIUM
RETICULUM
ERIDANUS
CAELUM
WEST
SOUTH
EAST
Horizon 20°N
Horizon 40°N
Horizon 60°N
40°N
20°N
Zenith

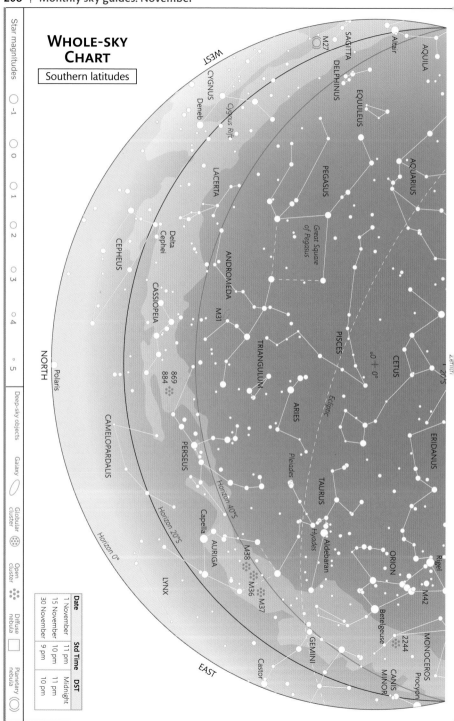

WHOLE-SKY CHART

Southern latitudes

Star magnitudes

○ -1
○ 0
○ 1
○ 2
○ 3
○ 4
° 5

Deep-sky objects

Galaxy ⬭

Globular cluster ⬤

Open cluster ⋮⋮

Diffuse nebula ▢

Planetary nebula ◎

WEST

NORTH

EAST

CYGNUS
Deneb
Cygnus Rift
LACERTA
Delta Cephei
CEPHEUS
Polaris
CASSIOPEIA
CAMELOPARDALIS
M31
ANDROMEDA
869
884
TRIANGULUM
PERSEUS
ARIES
Capella
AURIGA
LYNX
M38
M36
M37
Castor
GEMINI
M27
SAGITTA
DELPHINUS
Altair
AQUILA
EQUULEUS
AQUARIUS
PEGASUS
Great Square of Pegasus
PISCES
CETUS
Ecliptic
Pleiades
TAURUS
Hyades
Aldebaran
ORION
Betelgeuse
2244
MONOCEROS
Procyon
CANIS MINOR
Rigel
M42
ERIDANUS
Zenith
+0°
20°S
Horizon 40°S
Horizon 20°S
Horizon 0°

Date	Std Time	DST
1 November	11 pm	Midnight
15 November	10 pm	11 pm
30 November	9 pm	10 pm

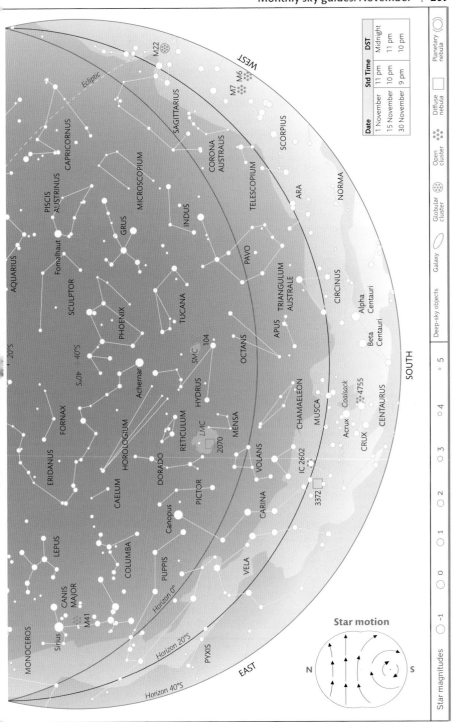

Date	Std Time	DST
1 November	11 pm	Midnight
15 November	10 pm	11 pm
30 November	9 pm	10 pm

WEST

SOUTH

EAST

Star motion

N — S

Star magnitudes

○-1 ○0 ○1 ○2 ○3 ○4 ○5

Deep-sky objects

Galaxy Globular cluster Open cluster Diffuse nebula Planetary nebula

DECEMBER

NIGHTS ARE AT THEIR LONGEST and days at their shortest in the northern hemisphere, and vice versa in the south. Much of the sky as it appears in the far south is occupied by constellations that were never seen by the astronomers of ancient Greece. A bright meteor shower radiates from Gemini in mid-month.

SUNRISE AND SUNSET ON 15 DECEMBER

Latitude	Sunrise	Sunset
60°N	09.00	14.50
40°N	07.10	16.40
20°N	06.30	17.20
0°	05.50	18.00
20°S	05.20	18.40
40°S	04.30	19.20

1 December: Std time ⊙ DST ⊙ | 15 December: Std time ⊙ DST ⊙ | 30 December: Std time ⊙ DST ⊙

NORTHERN LATITUDES

LOOKING NORTH Perseus is almost overhead with Capella and the other stars of Auriga to its right. Gemini is further east, with Cancer rising beneath it. On the northeastern horizon, Regulus leads the stars of Leo into view. In the northwest are Cassiopeia and Cepheus, with Deneb closer to the horizon.

LOOKING SOUTH Taurus is high in the south, ideally placed for observing the Pleiades and Hyades clusters (see p.213). Eridanus is beneath it, while to its lower left Orion is still rising, followed by brilliant Sirius and Procyon. Cetus can be seen in the southwestern sky, with Pisces and Pegasus visible progressively further to the west.

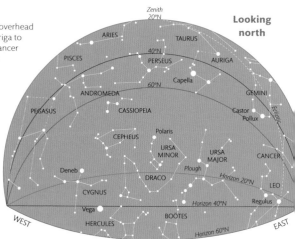

SOUTHERN LATITUDES

LOOKING NORTH Eridanus, the celestial river, meanders overhead. Taurus is below it, with the Hyades and Pleiades (see p.214) ideally placed and Orion high to its right. Auriga and Perseus are near the northern horizon, with Gemini in the northeast. In the northwest, the Great Square of Pegasus is setting, followed by Pisces.

LOOKING SOUTH Achernar and Canopus lie to the right and left of centre. In between them, slightly lower down, are the Large and Small Magellanic Clouds (LMC and SMC, see p.215). Sirius is high in the eastern sky, and Vela and Puppis are rising in the southeast. Fomalhaut is setting in the southwest.

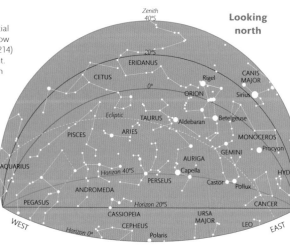

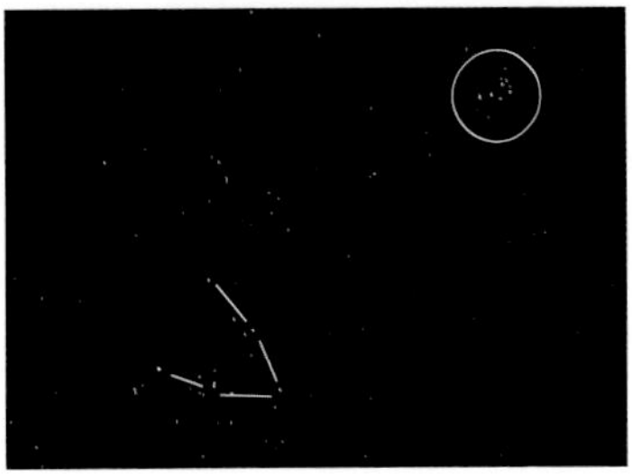

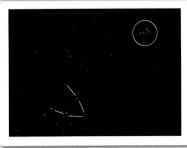

The Hyades and Pleiades (Taurus)

The stars of Taurus depict the top half of a bull's body. The large, V-shaped Hyades cluster (to lower left) outlines the face, and the red star Aldebaran represents one eye. The Pleiades cluster (top right) lies on the bull's back. See also panel, right.

Star magnitudes ○ -1 ○ 0 ○ 1 ○ 2 ◦ 3

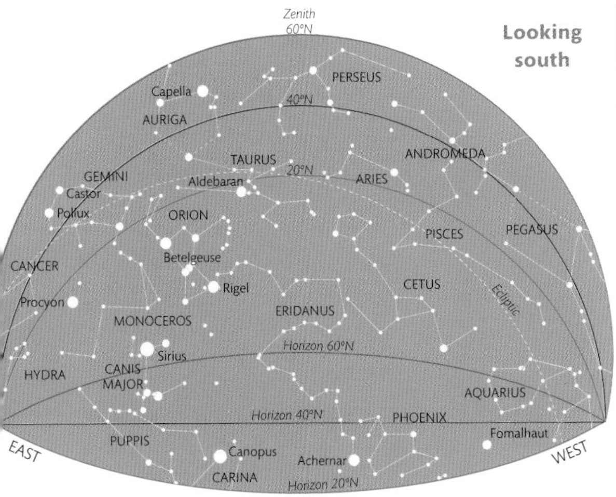

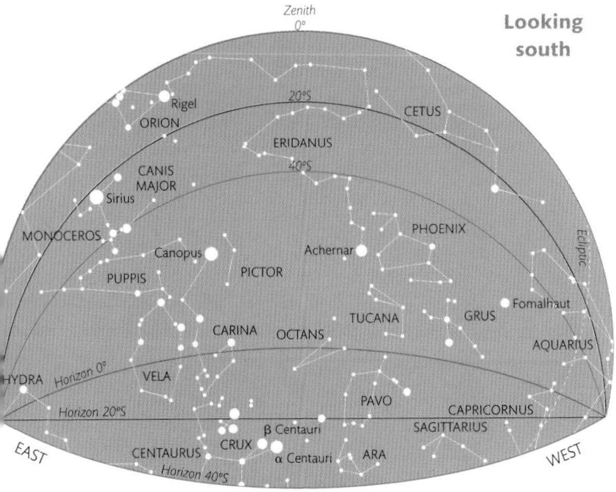

WHOLE-SKY CHARTS >>

December features

⁘ **M36, M37, and M38 (Auriga).** In the Milky Way star fields of Auriga, a crooked line of three open clusters can be seen through wide-field binoculars. M37 is the largest of them, while M36, in the middle of the line, is the smallest but also the easiest to resolve into individual stars with a small telescope. See also p.69.

• **The Hyades (Taurus).** This large, bright open cluster, the closest such cluster to Earth, contains more than a dozen stars visible to the naked eye. See also pp.132–33.

• **The Pleiades (Taurus).** At first glance, this open cluster appears as a hazy cloud, but a closer look reveals several individual stars. Although the cluster is also called the Seven Sisters, normal eyesight shows only about six stars. People with exceptional vision can see several more, and binoculars bring dozens of stars into view. See also pp.132–33.

Also visible

- ⁘ **M41** (p.151)
- ☐ **M42** (p.145)
- ⁘ **M44** (p.157)
- ☐ **NGC 2070** (p.205)
- ⁘ **NGC 2244** (p.151)

December meteors

The Geminids. The second-best shower of the year radiates from near Castor, in Gemini, reaching a maximum on 13 December, when as many as 100 bright meteors an hour may be seen. Lower rates of activity occur for several days either side of the maximum.

WHOLE-SKY CHART

Northern latitudes

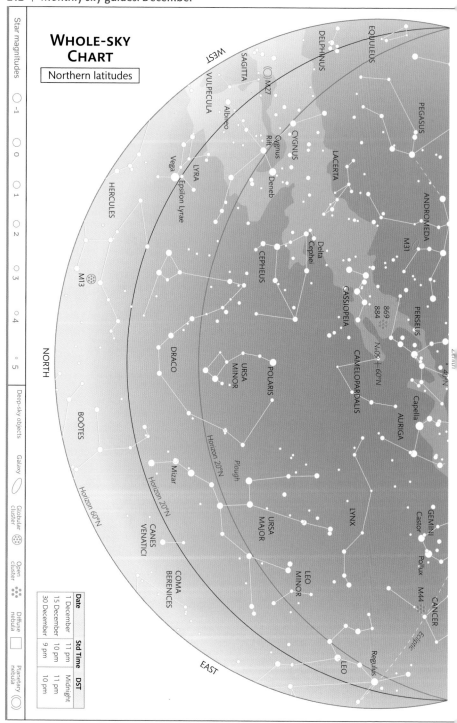

Star magnitudes

○ -1
○ 0
○ 1
○ 2
○ 3
○ 4
○ 5

NORTH

Deep-sky objects

Galaxy ◯

Globular cluster ◉

Open cluster ◌

Diffuse nebula ▢

Planetary nebula ◍

Date	Std Time	DST
1 December	11 pm	Midnight
15 December	10 pm	11 pm
30 December	9 pm	10 pm

WEST

EQUULEUS
DELPHINUS
PEGASUS
LACERTA
ANDROMEDA
M31
SAGITTA
VULPECULA
M27
Albireo
Cygnus Rift
CYGNUS
Deneb
Vega
LYRA
Epsilon Lyrae
HERCULES
Delta Cephei
CEPHEUS
CASSIOPEIA
869
884
PERSEUS
N.o9+62°N
Capella
AURIGA
CAMELOPARDALIS
Zenith
40°N
M13
DRACO
URSA MINOR
POLARIS
BOÖTES
Horizon 20°N
Plough
LYNX
Mizar
Horizon 20°N
URSA MAJOR
CANES VENATICI
COMA BERENICES
LEO MINOR
Horizon 60°N
GEMINI
Castor
Pollux
M44
CANCER
Ecliptic
Regulus
LEO
EAST

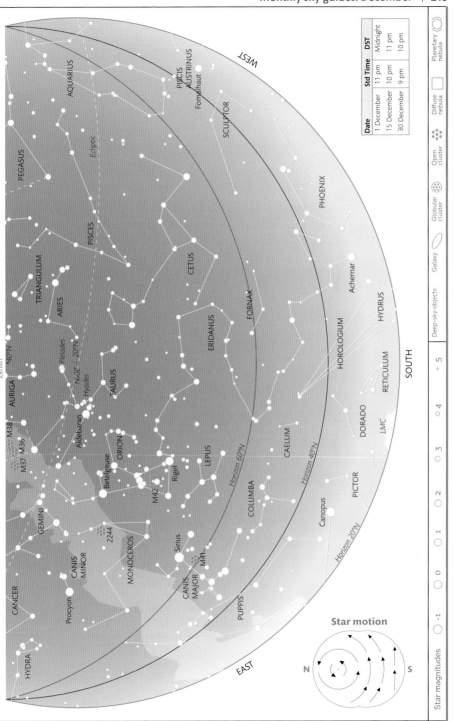

WEST

PISCIS
AUSTRINUS
Fomalhaut

SCULPTOR

AQUARIUS

PEGASUS

Ecliptic

PHOENIX

PISCES

TRIANGULUM

CETUS

ARIES

Achernar

FORNAX

HYDRUS

40°N

Pleiades
N20°Z
Hyades

ERIDANUS

HOROLOGIUM

AURIGA

TAURUS

RETICULUM

M38

Aldebaran

DORADO

M37 M36

CAELUM

LMC

ORION

Betelgeuse

LEPUS

SOUTH

Horizon 60°N

GEMINI

M42
Rigel

COLUMBA

Horizon 40°N

PICTOR

CANIS
MINOR

2244

MONOCEROS

Canopus

Procyon

Sirius

CANCER

M41

CANIS
MAJOR

Horizon 20°N

PUPPIS

HYDRA

EAST

Star motion

N   S

Date	Std Time	DST
1 December	11 pm	Midnight
15 December	10 pm	11 pm
30 December	9 pm	10 pm

Deep-sky objects

Galaxy

Globular
cluster

Open
cluster

Diffuse
nebula

Planetary
nebula

Star magnitudes

○ -1   ○ 0   ○ 1   ○ 2   ○ 3   ○ 4   ○ 5

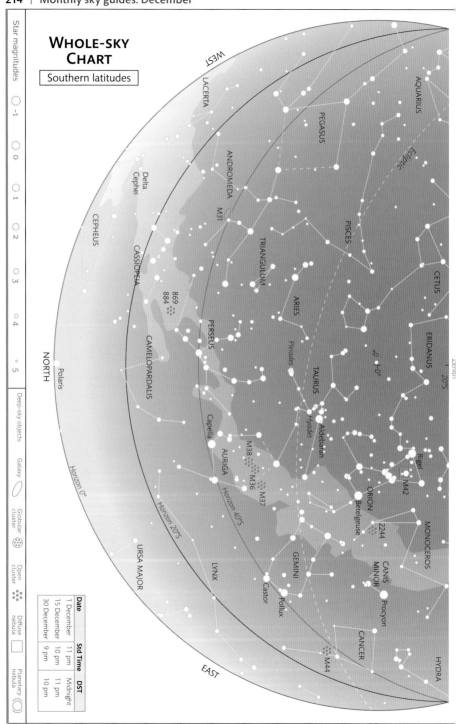

WHOLE-SKY CHART

Southern latitudes

WEST

NORTH

EAST

Star magnitudes

○ -1
○ 0
○ 1
○ 2
○ 3
○ 4
° 5

Deep-sky objects

Galaxy ⬭

Globular cluster ⬭

Open cluster ⬭

Diffuse nebula ☐

Planetary nebula ◎

Date	Std Time	DST
1 December	11 pm	Midnight
15 December	10 pm	11 pm
30 December	9 pm	10 pm

LACERTA

AQUARIUS

PEGASUS

ANDROMEDA

M31

Delta Cephei

CEPHEUS

CASSIOPEIA

PISCES

CETUS

TRIANGULUM

ARIES

ERIDANUS

869
884

PERSEUS

Pleiades

Ecliptic

CAMELOPARDALIS

TAURUS

Hyades

Aldebaran

20°S

0°

0°

Polaris

Capella

AURIGA

M38
M36
M37

Rigel

M42

ORION

Betelgeuse

MONOCEROS

2244

Horizon 0°

URSA MAJOR

Horizon 20°S

Horizon 40°S

LYNX

GEMINI

Castor

Pollux

CANIS MINOR

Procyon

CANCER

M44

HYDRA

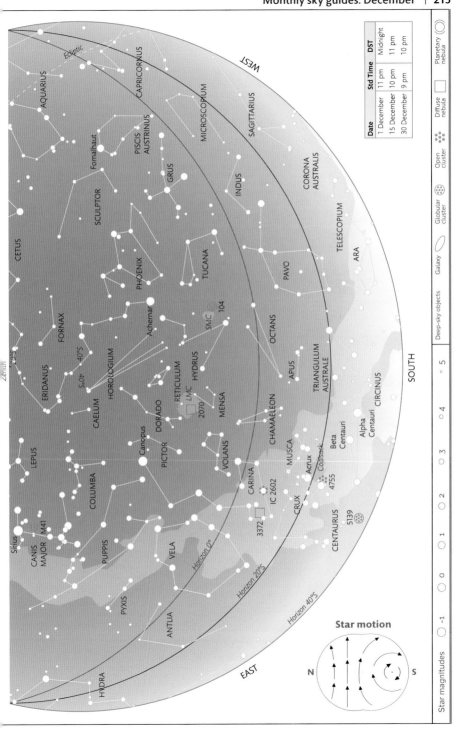

Date	Std Time	DST
1 December	11 pm	Midnight
15 December	10 pm	11 pm
30 December	9 pm	10 pm

Star motion

Deep-sky objects: Galaxy ◯ | Globular cluster ✳ | Open cluster ⁙ | Diffuse nebula ▢ | Planetary nebula ◯

Star magnitudes: ◯ -1 ◯ 0 ◯ 1 ◯ 2 ◯ 3 ◯ 4 ∘ 5

GLOSSARY

WITH THE EXCEPTION of words in regular use, words that are defined elsewhere in the glossary appear in bold type.

■ Absolute magnitude
A measure of the true **luminosity** of an object, defined as how bright it would appear if it were at a standard distance from the Earth, chosen to be 32.6 light years.

■ Aperture
The diameter of the light-collecting lens or mirror of an astronomical instrument.

■ Apparent magnitude
The brightness of an object as seen from the Earth. It is affected by the object's distance from us.

■ Asterism
A pattern formed by stars in one or more **constellations**. Examples are the Plough (or Big Dipper), which is part of Ursa Major, and the Great Square of Pegasus, formed by stars in both Pegasus and Andromeda.

■ Asteroid
A small, rocky body, also called a minor planet, orbiting the Sun.

■ Axis
An imaginary line about which a **celestial object** rotates.

■ Binary star
A pair of stars linked by gravity, orbiting around a common centre of mass. See also **Eclipsing binary, Spectroscopic binary**.

■ Black hole
A volume of space in which gravity is so great that nothing can escape, not even light, although objects can fall in. Black holes are thought to form when massive stars collapse.

■ Celestial object
An object in space, such as a star, planet, or **galaxy**, that appears in the Earth's sky.

■ Celestial equator
The celestial equivalent of the Earth's equator, lying on the **celestial sphere** directly over the Earth's equator.

■ Celestial poles
The celestial equivalent of the Earth's poles, around which the **celestial sphere** appears to turn each day.

■ Celestial sphere
An imaginary sphere of infinite size around the Earth, on which **celestial objects** appear to lie. See also pp.16–17.

■ Cepheid variable
A type of **variable star** that pulsates regularly in size, every few days or weeks, with a period that is linked to its average **luminosity**.

■ Circumpolar star
A star that is above the horizon all night, as seen from a particular location on the Earth. Instead of rising and setting, the star appears to circle one of the **celestial poles**.

■ Comet
A ball of frozen gas and dust on a long orbit around the Sun.

■ Conjunction
An occasion when two bodies in the **Solar System** (usually the Sun and a planet) line up as seen from the Earth. See also **Inferior conjunction, Superior conjunction**.

■ Constellation
An area of sky, originally a star pattern, but now defined as the area within boundaries set by the International Astronomical Union.

■ Declination (dec.)
The angle between an object and the **celestial equator**. Declination is the equivalent of latitude on the Earth. It is measured at right angles to the celestial equator, from 0° on the equator itself to 90° north or south at the **celestial poles**.

■ Deep-sky object
A term applied to star clusters, **nebulae**, and galaxies.

■ Diffuse nebula
A bright cloud of gas, illuminated by stars within it.

■ Double star
Two stars that look close together as seen from the Earth. See also **Binary star, Optical double star**.

■ Dwarf planet
An object orbiting the Sun that is spherical in shape, or nearly so, but is not large enough to have cleared other objects out of its orbit.

■ Eclipse
An occasion when one **celestial object** cuts off some or all of the light from another.

■ Eclipsing binary
A pair of stars in orbit around each other, in which one star periodically passes in front of the other as seen from the Earth, cutting off its light.

■ Ecliptic
The plane of the Earth's orbit around the Sun, projected on to the **celestial sphere**. The Sun appears to move along the ecliptic, but this is in fact due to the Earth's motion around the Sun.

■ Elongation
The angle between a planet and the Sun, or between a moon and a planet, as seen from the Earth.

■ Equinox
The occasion when the Sun lies on the **celestial equator**. This occurs twice a year, around 21 March and 23 September. At an equinox, day and night are roughly equal in length everywhere on the Earth.

■ Galaxy
A mass of stars, numbering from millions to billions, all linked by gravity. Galaxies range from about a thousand light years to hundreds of thousands of light years across.

■ Giant star
A star that has become bigger and brighter in the late stages of its evolution.

■ Globular cluster
A ball-shaped group of stars, about 100 light years across, with tens of thousands to hundreds of thousands of members. Such clusters contain some of the oldest stars known.

■ Inferior conjunction
The occasion when either Mercury or Venus lies between the Sun and the Earth.

■ Light year
A unit of distance. It is the distance a beam of light covers in a calendar year, or 9,460,700,000,000km (5,878,600,000,000 miles).

■ Local Group
The cluster of over 50 galaxies that includes our own.

■ Luminosity
The inherent brightness of a light-producing **celestial object**.

■ Magnitude
The brightness of a **celestial object**, measured on a numerical scale on which bright objects are given low or negative values and faint objects have high values. The brightest stars are termed first magnitude, fainter ones second magnitude, and so on. In general usage, an object described as being of, for example, fourth magnitude

has a magnitude between 3.5 and 4.49. *See also* **Absolute magnitude**, **Apparent magnitude**. *See also* p.18.

■ **Meteor**
A streak of light in the sky, caused by a small particle burning up in the Earth's atmosphere.

■ **Meteorite**
A fragment of an **asteroid** that has fallen to the surface of a planet or its satellite.

■ **Milky Way**
The hazy band that can be seen in the sky on dark nights, composed of billions of distant stars in our own **Galaxy**. It is also a popular name for our Galaxy as a whole.

■ **Mira variable**
A red **giant** or **supergiant star** that pulsates in size over a period of months or years, varying in brightness by as much as 11 **magnitudes**.

■ **Nebula**
A cloud of gas and dust, usually found in the spiral arms of a **galaxy**. *See also* **Planetary nebula**, **Diffuse nebula**.

■ **Neutron star**
A small and highly dense star consisting of the atomic particles known as neutrons, and thought to be the remains of a massive star that has exploded as a **supernova**.

■ **Nova**
A star that erupts temporarily in brightness by many thousands of times. Novae occur in close **binary stars**, one member of which is a **white dwarf**. Gas flows from the companion **star** on to the white dwarf, causing a small explosion which leads to the brightening.

■ **Open cluster**
An irregularly shaped group of dozens or hundreds of relatively young stars, usually found in the spiral arms of a **galaxy**.

■ **Opposition**
The occasion when a body in the **Solar System** appears opposite the Sun, as seen from the Earth, and is hence visible all night.

■ **Optical double star**
Two stars that appear close to one another in the sky but in fact lie at different distances from the Earth.

■ **Orbit**
The path of an object that is moving through space under the gravitational influence of another body of greater mass.

■ **Parallax**
The change in position of an object when seen from two different locations. How much the object's position changes depends on its distance and the separation of the observing locations. Nearby stars show a slight parallax shift as the Earth orbits the Sun, from which their distances can be calculated.

■ **Phase**
A fraction of the illuminated disc of a planet or moon, as seen from the Earth.

■ **Planet**
A relatively large object that orbits a star and does not emit light.

■ **Planetary nebula**
A shell of gas thrown off by a star late in its development.

■ **Radiant**
The point from which the meteors in a shower appear to diverge.

■ **Reflecting telescope**
A telescope in which light is collected and focused by a mirror.

■ **Refracting telescope**
A telescope in which light is collected and focused by a lens.

■ **Retrograde motion**
Movement from east to west (or clockwise as seen from above an object's north pole). This is opposite to the general direction of motion in the **Solar System**.

■ **Right ascension (RA)**
A coordinate on the **celestial sphere**, the equivalent of longitude on Earth. It is measured in hours (one hour being equivalent to 15 degrees) from the point where the Sun crosses the **celestial equator** every March (the vernal **equinox**).

■ **Solar System**
The Sun and the various bodies in orbit around it. These include the eight planets and their moons, as well as **dwarf planets**, asteroids, comets, and smaller pieces of debris.

■ **Solstice**
The occasion when the Sun reaches its furthest point either north or south of the **celestial equator**. This happens around 21 June, the longest day in the northern hemisphere, and 22 December, the longest day in the southern hemisphere.

■ **Spectroscopic binary**
A pair of stars so close together that they cannot be separated with any telescope. The fact that the star is a

binary is revealed only when its light is studied with a spectroscope.

■ **Star**
A ball of gas that generates heat and light as a result of nuclear reactions at its core.

■ **Supergiant star**
A star at least ten times as massive as the Sun, and which has become bigger and brighter in the late stages of its development. Supergiants can be hundreds of times the diameter of the Sun, and tens of thousands of times as bright.

■ **Superior conjunction**
The occasion when Mercury or Venus lies on the far side of the Sun, as seen from the Earth.

■ **Supernova**
An explosion in which, for a few weeks or months, a star brightens by millions of times. Only the most massive stars become supernovae.

■ **Universe**
Everything that exists, including all matter, space, and time.

■ **Variable star**
Any star that appears to change in brightness. This is usually due to pulsations in the size of the star, but some variables are actually close **binary stars** in which one star periodically **eclipses** the other. *See also* **Cepheid variable**, **Mira variable**.

■ **White dwarf**
A small, dense star with a mass similar to that of the Sun, but only about 1 per cent of the Sun's diameter.

■ **Zenith**
The point in the sky directly above an observer.

■ **Zodiac**
The band of sky either side of the **ecliptic** through which the Sun and planets move. Although there are 12 **constellations** of the zodiac dating from ancient times, more recent changes in constellation boundaries mean that the Sun and planets also pass through a 13th constellation, Ophiuchus.

INDEX

ACKNOWLEDGEMENTS

THE AUTHOR AND PUBLISHER would like to thank the following people for their invaluable help: Robin Scagell for advice on photography and telescopes; and the staff at Her Majesty's Nautical Almanac Office, (particularly Steve Bell, David Harper, Catherine Hohenkerk, and Andrew Sinclair) for their work on the star charts and planet location diagrams.

DORLING KINDERSLEY would also like to thank: Lynne Marie Stockman for reference material for the inset chart of the Pleiades on p.133; Giles Sparrow for checking the constellation figure artworks; and Lesley Riley for editorial assistance in the first edition, and for this edition: Mayank Shankar Choudhary for picture research; Anjali Sachar and Mridushmita Bose for design assistance; Kathakali Banerjee and Radhika Haswant for editorial assistance.

ILLUSTRATION CREDITS

All star charts and planet-location diagrams plotted by Her Majesty's Nautical Almanac Office. Solar System locator illustrations by Rob Campbell. Illustrations of the internal structure, and the rings and moons of the gas planets by Luciano Corbella. Tilt, rotation, and orbit illustrations and pie-charts of atmospheric composition by Martin Cropper. Other illustrations are by the following artists/organisations: **Julian Baum:** 14 (b), 26 (all), 29 (c), 32 (c), 38, 41 (c), 42 (cl & cr), 51, 54 (c), 56 (c), 59 (t), 60 (t), 61; **Richard Bonson:** 12 (c), 13 (all), 14 (t & c), 19 (bl), 23 (all); **Rob Campbell:** 12 (b), 27 (c); **Luciano Corbella:** 9 (b), 19 (br), 34 (cl), 44 (cr); **DK Cartography/Planetary Visions Ltd (London):** 36 (c); **DK Cartography:** 142 (t); **John Egan:** 16 (all), 17 (all), 53, 60 (bl); **Helen Taylor:** 15 (tr, bl, & br), 18 (b), 20 (b), 21 (all), 33 (c), 36 (b), 39 (c), 43 (c), 44 (cl), 47 (t), 59 (c); **King and King:** 28 (all), 62.

PHOTOGRAPHY CREDITS

The following pictures were provided by Galaxy Picture Library: **Robin Scagell:** 27 (clb), 30 (tr), 33 (tr), 43 (tc & tr), 47 (cl), 51 (tr), 60, 64, 67 (both), 68, 69, 70, 78, 81, 85, 87 (tl & br), 88 (br), 90 (cr), 99, 103, 105, 121, 124, 126, 133, 137, 138 (both), 145, 163, 169, 175, 187, 199, 211; **Stan Armstrong:** 28 (tr, cr & tc); **Stephen Fielding:** 129; **Chris Floyd:** 127 (br); **Bob Garner:** 43 (c), 47, 51 (cr); **Gordon Garradd:** 74, 76, 106, 114, 122, 127 (bl), 134, 151; **Maurice Gavin:** 33 (c), 41 (b), 47 (cr), 51 (c); **Eric Hutton:** 115 (tl inset); **JPL:** 35 (all), 45 (cl & cr), 48, 49 (all); **Chris Livingstone:** 205, 193; **Rob McNaught:** 181; **Brian Manning:** 90 (br); **Michael Maunder:** 28 (br); **NASA:** 53 (br), 311 (Calvin J. Hamilton), 41 (l), 53 (cl), 57; **Terry Platt:** 137 (tr); **Pedro Re:** 99 (tr), 130, 140; **Royal Observatory, Edinburgh and Anglo-Australian Observatory (©) 1982:** 10 (both); **Trevor Searle & Phil Hodgins:** 137 (bl); **Charles Smith:** 22 (br);

Michael Stecker: 64, 66, 125 (all), 127 (tl), 128, 131; **Nick Szymanek/Ian King:** 83, 120, 132, 137 (br); **STScI:** 10, 59, 131 (tr).

Pictures from other sources: **Anglo-Australian Observatory:** 9 (all), 13 (c), 135; **Corbis:** Roger Ressmeyer: 8–9 (t); **Andy Crawford:** 20 (all), 21, 143; **Dave King:** 22 (cr); **Hale Observatory:** 27 (tr); **Alamy Stock Photo:** astropix 73 (crb), Azoor Travel Photo 6 (c), Galaxy / Robin Scagell 95 (tr), Steven Milne 115 (crb), Stocktrek Images / Alan Dyer 157 (tl); **© Altair Astro Limited 2022:** 24 (br); **Adam Block:** Mount Lemmon SkyCenter / University of Arizona 107 (crb); **Bob Franke:** 72 (crb); **Image courtesy of Celestron:** 23 (all); **David Cortner:** 40 (cb); **Dorling Kindersley:** NASA 28 (cra), 40 (phases of the moon); **Dreamstime.com:** Antonio Corrado 22 (bc), Tognarini Franco 96 (br), Gryzeva 2, Johnnydevil 17 (tr), Bogdan Lazar 115 (tl), Scol22 22 (Binoculars) (br), Alexandr Yurtchenko 93 (tl); **ESO:** 79 (cr), 116 (crb), 139 (crb), S. Brunier 87 (tr), TRAPPIST / E. Jehin 100 (cl); **Fotolia:** Leonid Smirnov 24 (cl); **Robert Gendler:** 91 (cr), 110 (crb); **Getty Images:** 500px / Alessio Beltrame 89 (tr), imagenavi 77 (bl); **Getty Images / iStock:** Manfred_Konrad / E+ 141 (cr), Kevin Morefield 111 (tr); **Euan Mason:** 94 (bl); **NASA:** Bill Ingalls 30 (cla), ESA 108 (br), ESA / Hubble 113 (cl), ESA, A. Simon (Goddard Space Flight Center), M.H. Wong (University of California, Berkeley), and the OPAL Team 50 (c), ESA, and M. Brown (California Institute of Technology) 59 (br), ESA, Hubble 117 (clb), ESA, STScI, A. Simon (Goddard Space Flight Center), and M.H. Wong (University of California, Berkeley) and the OPAL team 46 (c), GSFC / Arizona State University 39 (cb), JHUAPL / SwRI 58 (c), Johns Hopkins University Applied Physics Laboratory / Carnegie Institution of Washington 31 (br), JPL 62 (cra), JPL-Caltech / MSSS 45 (cb), JPL-Caltech / SwRI / MSSS / Kevin M. Gill 48 (crb); **Chris and Dawn Schur/Cybertrails.com:** 97 (cla); **Science Photo Library:** NASA/GSFC 37, ROBIN SCAGELL 108 (bl), US NAVAL OBSERVATORY 89 (bl); **Shutterstock.com:** Brian Donovan 77 (tr), 86 (crb), Manos Malakopoulos 13 (tr), Heiner Weiss 118 (crb); **Sky-Watcher Copyright © 2011–2018 Pacific Telescope Corp. All rights reserved.:** 22 (c), 24 (c); **Suzhou ZWO Co., Ltd.:** 24 (cb); **Peter Wienerroither:** 12 (ca), 88 (br inset); **Endpaper images:** Front & Back: **123RF.com:** rustyphil; **Dreamstime.com:** Martin Holverda, **Science Photo Library:** NASA.

All other images © Dorling Kindersley Limited